All in a Summer's Fun...

JEFF KING: A handsome hustler with cobalt-blue eyes. He came to the Cape to score easy money...and made the mistake of cashing in on a deadly debt.

ANN RICHARDSON: A pretty Boston girl happy in a summer love. Her love of drugs could turn this summer into a lethal affair.

NOAH LOVELACE: Clarisse's wealthy uncle. The family has nothing to say about how he spends his money...but somebody is interested in whom he plans to leave it to.

THE WHITE PRINCE: Noah's spoiled lover. The party with Noah is ending...and he's not likely to end the relationship as friends.

TERRY O'SULLIVAN: A Boston publishing executive. He has a bad case for Dan Valentine...and a secret that could prove fatal to tell.

Old friends Dan Valentine and Clarisse Lovelace are forced to include investigation on their vacation schedule when Provincetown becomes the scene for intrigue and murder.

Other Avon Books by
Nathan Aldyne

VERMILION

Coming Soon

SLATE

COBALT

Nathan Aldyne

AVON
PUBLISHERS OF BARD, CAMELOT, DISCUS AND FLARE BOOKS

AVON BOOKS
A division of
The Hearst Corporation
959 Eighth Avenue
New York, New York 10019

The St. Martin's Press edition contains the following Library of
Congress Cataloging in Publication Data:

Aldyne, Nathan.
Cobalt.

I. Title.
PS3551.L346C6   813'.54   81-21540
                                  AACR2

First Avon Printing, October, 1982

For Sharon

Prologue

Wednesday, 3:00 P.M.:

"Hello. This is not Daniel Valentine, but his disembodied voice brought to you through the miracle of printed circuitry. When you hear the beep tone, please leave your name, number, and a short message. When I return I will erase the tape. Thank you."

"It's Clarisse at three o'clock. Life or death."

Wednesday, 7:00 P.M.:

"Hello?"

"Whose life, whose death?"

"Thank God you've called, Val!"

"Whose life, whose death?"

"Veronica Lake."

"She's dead? Oh, Clarisse, I'm so sorry, I—"

"She's not dead, she just threw up in the fireplace."

"I don't know how you expect to keep a dog in the city anyway. Why did you really call?"

"I was feeling sorry for myself. When Veronica Lake got sick I took her out to my brother's, and now I'm *completely* alone. I wish I were with you in Provincetown! Boston is awful. The Esplanade is *not* Herring Cove, Newbury is *not* Commercial Street, the city is filled with tourists who are apparently seeing concrete sidewalks for the first time, the health club has closed the pool for renovations, and the rental season is *unmercifully* slow."

1

"Hate to say this, but the rest of the summer isn't going to be any better if you insist on staying in Boston. Why don't you come down to Provincetown? Quit your job, pack your bags, lock up the flat, and hop on the ferry."

"The utter stupidity of that idea has considerable appeal for me. But, Val, where would I stay?"

"You can stay with me."

"You just want somebody to help with the rent. And you forget, I wouldn't have any income down there. I'd need a job, I'd—"

"I've already got you a job."

"Doing what?"

"There's a shop right across from the bar where they need somebody, and I've already mentioned you."

"What sort of place is it?"

"Gift shop—rare and beautiful things. Come on, Clarisse, I guarantee that you will be completely happy here. Absolutely nothing will go wrong the entire summer. Sun, fun, and romance—that's what life is like in P'town. And besides, the first big party of the season is Saturday night. Private. Invitation only. Open bar."

"Who's giving it?"

"The Crown. And of course, this being Provincetown, there's a theme."

"What is it?"

"Garden of Evil."

"It'd take me two days to get up a decent costume."

"So you'll come?"

"If you promise to get rid of the answering machine."

"Hear that sound? It's me, ripping the plug from the wall. I'll see you Saturday then?"

"Oh, why not? Tell Noah I'm coming, and put some clean sheets on the bed. Oh and by the way, how much does that job pay?"

"A third of what you make now. Poverty will stare you in the face."

"I don't care. I'll take it. I have to be in P'town. Valentine, I love you. It's a real problem."

"Come live with me then. That'll get you over it."

PART I
The Garden of Evil

Chapter 1

At a quarter past one the following Saturday afternoon as the Provincetown ferry was being secured to the wharf, Clarisse Lovelace, attired in a white sailor-suit top with blue piping and matching white bell-bottomed pants, was first in line to disembark. A little girl had tried to slip in front of her, but when Clarisse pointedly remarked that it wasn't too late to be hurled overboard the child retreated. The ankle strap of one of Clarisse's heavy-heeled sandals was loosened to lessen the pressure on a large blister that had developed on her heel since the morning, her cascade of black hair was tangled about her shoulders from having been whipped by the salt wind for the past three hours, and her oversized octagonal sunglasses were perched awry on her nose—the right-hand stem had been broken by an ecology freak rushing to the railing when whales were sighted off the port bow. Her sailor's cap had blown off before the ferry had even left Boston harbor. Clarisse's back ached from carrying her overstuffed leather bag, and when she hoisted it over her shoulder, a thick strand of her hair caught in the zipper.

"Move it, lady!" urged the three dozen or so day-trippers directly behind her, who were desperately eager to trample Provincetown in the three hours they had before the ferry began its voyage back to Boston.

She turned with a glance of loathing for them all.

When she reached the wharf, she stepped quickly to one side. As she painfully disentangled her hair from the zipper of

5

her leather bag, she watched her fellow passengers swarming off the ferry. The travel bag was dropped onto the rough weathered boards and the costume for that night's party, in a suit bag, laid carefully over it. Behind her a gaggle of adolescent boys in swim trunks and diving goggles were splashing in the water, shouting "Coins, coins!" up at the passengers. Several amused women stopped to toss pennies, but the divers contemptuously allowed these to sink, and screeched "Quarters! Throw some quarters!"

The teenaged voices had anything but a salutary effect on Clarisse's headache. She stepped to the edge of the wharf and, when one boy whose voice was particularly harsh shouted, "Throw, throw!" Clarisse ripped off her broken sunglasses and flung them at his head.

She picked up her bag and moved down the long pier. Before her, Provincetown was spread in a multicolored crescent along the inside of the Cape Cod hook. As she trudged along with her bags she watched eagerly for a sight of Daniel Valentine, but saw neither his face nor form. One of the very few handsome men she had seen on the ferry moved along beside her almost in step. He was of medium height and size, and much more than medium good looks, with short dark hair and a carefully trimmed mustache. His skin was flawless and though the summer was just under way, already well tanned. He wore black sneakers and button-fly jeans. His shirt dated from the fifties: bright red, patterned in lines of small black tulips, with the long sleeves carefully rolled to encircle his large biceps. But it was his eyes that most drew Clarisse's attention: they were a startling cobalt blue. When Clarisse paused, exhausted, he stopped and offered to carry her bag for her. She accepted gratefully.

"My name's Jeff," he said, then amended, "Jeff King."

"I'm Clarisse," she replied, but did not offer her last name.

"Are you down for the weekend?"

"No, I'm here for the summer. But the fact is," she added confidingly, "I *hate* resorts."

"Where are you staying?" Jeff asked.

"At my uncle's place."

"You're lucky. I tried to get a reservation, but there wasn't anything available. I'll have to see what turns up."

6

Clarisse looked him over and laughed. "I imagine you'll come across someone with an extra pillow."

Jeff smiled at the compliment. "I hope so. I used to come down here a lot, and I had a lot of friends here. I guess I'll have to see who's in town this season."

They had reached the municipal parking lot, and Clarisse thanked Jeff for his assistance.

"Where does your uncle live? I'm not doing anything, I might as well take it on for you."

Clarisse, sensing that Jeff wanted nothing more in the world than for her to offer her uncle's living room couch as a place to stay the weekend, smiled warmly, and said, "Thank you, but a friend is supposed to be meeting me. Of course, if he's not here in five minutes, I have every intention of murdering him." She collapsed onto a piling that looked to have a tolerably clean surface. "I'm just going to sit here for a few minutes and put myself together. A woman resolved to commit a capital crime can't be too careful about her appearance."

"You look great," said Jeff. "I noticed you on the ferry. Your outfit looks great."

He seemed disposed to linger, perhaps to see if the compliment had assisted his cause, but Clarisse put her hand around the handle of her bag and politely wrested it from him. "Thank you again," she said in a tone of voice that did not brook argument.

After an awkward moment in which he swung his own bag to and fro, Jeff said, "There's a big costume party tonight."

"I know," replied Clarisse.

"Maybe I'll see you there," Jeff continued lamely.

"Of course. I'll be the one with blood on my hands." She pointedly turned her head toward the town, as if searching for her friend, and Jeff walked on.

Clarisse sighed, opened the zipper of her bag a few inches, and rummaged inside. When she found her brush, she pulled it violently through her hair until she thought it might be just presentable, and then stood and straightened the shoulders of her blouse. She opened her bag further and extracted a bottle of aspirin and gulped three down dry. She took out her box of adhesive bandages, and placed one over the blister on her heel. She stood, hoisted her bags with a groan, and set off for the Throne and Scepter.

The early afternoon was cloudy, and the brisk salt air was spiked with the scent of impending rain, but Clarisse was well enough acquainted with the unpredictability of Cape Cod weather to distrust her senses completely. In Provincetown you might *taste* rain, and still hope for a brilliantly sunny afternoon.

Saturday afternoon had brought a full complement of tourists to the town. Commercial Street, the principal thoroughfare, which follows the line of the bay and beach for the entire length of the crescent-shaped town, was lined on both sides with families from inland states, couples who doubtless thought themselves in love, and little knots of sullen teenagers who had been told that Provincetown was the hottest place on the Cape but now were at a loss to determine what raised the temperature so. The gay men and lesbians were either still in bed, already at work, or sitting at the Boatslip feeling guilty about starting to drink so early in the day. Turning onto Commercial Street, Clarisse pushed her way along the narrow sidewalk, constantly smiling and saying "Excuse me, please, I'm pregnant," until she found herself standing before the Throne and Scepter. It was half past one, but already the tiny tables placed among the green palms on the shaded veranda were taken up with chatting tourists who had turned their chairs so that they might watch the ceaseless parade along Commercial Street. A thin young man whose surliness qualified him for any waiter's job in Provincetown glanced with disdain at Clarisse and her baggage, but she ignored him and barged through the open French doors into the bar.

In the sudden dimness of the interior, she could barely make out more potted palms, lazily swirling ceiling fans, and mirrors set to catch the reflections of the street. Clarisse lurched forward to where she remembered the bar to be.

"Pour me a drink before you die," she gasped, and in another moment, as her vision began to take in more detail, she saw a glass of ice and clear liquid sitting on the bar before her.

"I've been waiting for you," said Daniel Valentine. His

blond hair and beard were lighter than when she'd last seen him, and much more closely cropped.

"I expected a deeper tan."

Valentine shrugged and automatically made a preening motion of tucking in his clinging red T-shirt. His sleek, tapered muscularity strained the cotton. "How can I get a tan when I've got the day shift? That's why I didn't meet you at the ferry. How *was* the ferry?"

"The ferry was the most horrible experience of my entire life," said Clarisse evenly. "It was insult, torture, and degradation." In three long swallows, she had finished her drink. She hadn't put the glass down before another took its place. Valentine drew a pack of Luckys from the back pocket of his jeans, lit two, and handed one to Clarisse.

"I've never ridden the ferry," he remarked. "I thought it was supposed to be quaint or something." He glanced at her sailor's outfit. "Did you keep getting mistaken for crew?"

"There was a Dixie Cup jazz band," said Clarisse. "It was amplified. And it played polkas. For three hours. A great number of people danced. They danced the polka. The people who didn't dance the polka got drunk and sang sentimental Irish songs. The people who didn't dance *or* sing, threw up. Oh yes, and on the upper deck, where I got to sunbathe for half an hour before the sun went behind a cloud, there was an eighty-one-year-old man who stood on his head and delivered a lecture on the dangers of tobacco."

"Did you meet anybody cute?"

"Well, there were approximately nine hundred persons on board the ship. I counted three attractive persons. Two women—very sweet and doing a duet of 'I Only Have Eyes for You.' And one man—who wanted me to put him up for the weekend."

"Sounds promising."

"He was gay—but I don't think he knew *I* knew that."

"Still sounds promising. Did you make an offer for me?"

"Valentine, I am very unhappy. My new sunglasses were torn off my head and smashed. A little boy *sat* on the costume that I had planned to wear tonight. I have a blister on my heel and a headache that only death will cure. I'm in no condition to pick up tricks for you."

"Well, you're here. That's something."

Outside, in the street, cars moved haltingly, trying to make headway through the milling throngs of pedestrians. A disgruntled driver blasted his horn at three women on roller skates who banged his trunk as they went by. A child shrieked when its ice cream cone was gobbled up by a passing mastiff. Someone wearing a large felt hat fashioned in the likeness of a goose peered in the window. The sun was suddenly obscured by a thick cloud, and there was a low bellow of thunder. Clarisse looked around the dark, hot, empty bar. "For this I quit my job, and sublet my rent-controlled apartment? Where's the sun?" she demanded. "Where's the fun? Where's the romance?"

Chapter 2

For a sum he could not bear to mention aloud, Daniel Valentine had rented for the summer one-third of the house that belonged to Clarisse's uncle. Four years before, Noah Lovelace had bought the low, rambling, U-shaped house on Kiley Court as an investment. He had broken it up into three fair-sized apartments that opened onto a central court with a swimming pool. In one of the apartments Noah lived with his companion of many years, a man called Victor, but more commonly known—especially in Provincetown—as the White Prince. In the apartment directly across the pool from Noah and the White Prince lived Valentine, and now with him, Clarisse. The third apartment, between Valentine's and Noah's, was rented out by the week.

Valentine sat in the twilit courtyard relaxing with a gin and tonic after his noon to eight o'clock shift at the Throne and Scepter. The storm that had threatened earlier swept out to sea, taking with it the humidity that had oppressed Clarisse on the ferry. The evening was clear and temperate. The sun had set and the sky was a luminous azure.

Valentine made a decent wage at the bar and consistently received generous tips—not just because he was efficient, which he was, nor simply because he was congenial and a good listener when the occasion warranted, which it often did, nor only because he was handsome and hot, but because of a smoothly balanced combination of all three of these elements. In Boston, Valentine had worked at a small bar in

Bay Village, but he had considered that job as temporary as the one he had now taken for the duration of the Province-town summer. Though he wouldn't admit it even in the most drunken confessional, Daniel Valentine was at heart a social worker. He had lost his job in the Suffolk County prison system, working with inmates shortly to be released, when he uncovered a scandal in the Sheriff's Department. The sheriff still held his job despite Valentine's revelation, in the pages of *The Real Paper*, that he had, with state funds, purchased ten thousand dollars' worth of crushed velvet draperies for his living room. Now, however, the sheriff was ailing, and his ailment looked to be terminal—this privileged information had been obtained from a radiologist at Massachusetts General Hospital who lusted for Valentine's embrace. Valentine hoped to regain his position at the Charles Street Jail by the autumn or early winter at the latest. He took a long swallow of his drink to toast the destruction of the sheriff's remaining leukocytes.

Valentine looked around with satisfaction. Twice before he had spent summers in Provincetown, but never had he lived in such congenial surroundings. The pool took up almost half the courtyard's area, and the house, covered in cedar shin-gling weathered a uniform gray, hugged close around it. Ivy rampaged over the walls, and multiflora roses of yellow and vermilion competed with the ivy in all the corners and around the doors. An enormous Kentucky coffee tree—the only one in Provincetown—stood just beside the latticed gate and its broad flat leaves sheltered the entire courtyard. The low flower beds were thick with lavender and nicotiana, plants that could take the shade and which made the whole place fragrant at night. Just beyond the latticed fence was Kiley Court, high-hedged, narrow and graveled. The only traffic here was the occasional provisions truck that scrunched its way down to the restaurant that was situated directly opposite Noah's compound.

While Valentine sat with his drink, windows had one by one lighted up in the three apartments opening onto the courtyard. If he looked to his left, he could see Clarisse wandering from room to room, looking for something she could not find, and waving her hands in frustration. Her cursing was a pleasant murmur, like the wind through the coffee tree. To his right, he could see into the bedroom of the White Prince. Directly beneath a harsh sunlamp sat Victor, his perfect vacuous face smeared over with gray cream and his proud white hair carefully protected by a skullcap made of crushed aluminum foil. While Valentine was peering into the lighted living room of the rented apartment in between, trying to glimpse the new tenant, Noah Lovelace's screen door slammed and Noah came out with a drink for himself and a freshened gin and tonic for Valentine.

"Have you seen the new ones yet?" he said in a low voice, nodding toward the rental apartment. Clarisse's uncle scraped a chair up next to Valentine. Noah Lovelace was in his late forties, though he looked no more than a weathered thirty-eight; he had short gray hair, a close-cut beard that was fast going to gray, and a body that was Valentine's good-natured envy. He was the only brother of Clarisse's father, and though rather a black sheep in the family, had managed to make more money than anyone else, having invested his niggardly inheritance in real estate and over-the-counter stocks with spectacular results. His disinclination to have any regular profession, his sexual orientation, his insufferable financial luck, and his insane decision to live year-round in a tourist-trap resort infuriated all the Lovelaces but Clarisse.

"No," said Valentine, "but I suppose anybody would be better than Terry."

Noah laughed, and raised his glass. "To the departure of Mr. O'Sullivan. I suppose he got on your nerves, always hanging about the way he did."

Valentine shrugged. "He's sweet, and he means well, but—"

"I know," replied Noah. "He's very—what should I say?—*sincere*. But he's left a representative."

"What?" asked Valentine apprehensively.

"His administrative assistant has the apartment this week. Her name is Ann, and her girlfriend's name is Margaret. They're high-tech lesbians, and they're in the first heat of love."

"How do you know that?"

"They've been going at it all afternoon. The Prince stood

13

with his ear against their bedroom wall for forty-five minutes, ticking off the orgasms on an abacus."

To their left a window suddenly shot up in its frame, and Clarisse's towel-turbaned head leaned out into the gathering night. "It's the end of the world!"

"What's wrong?" asked Valentine calmly.

"I forgot to bring my hair dryer! And I can't find yours!"

Valentine ran his hand over his head: his hair wasn't more than an eighth of an inch long. "I scrapped mine," he said.

"What am I supposed to do?"

"Try the oven."

"Victor has at least three dryers," said Noah. "Come on over, Clarisse, I'm almost certain he doesn't use more than one at a time."

The window shot down again, and a moment later Clarisse flung herself out the door. When she passed Valentine she grabbed the drink from his hand and guzzled half of it. "Thank you," she breathed. Over the rim she peered at him in the obscurity of twilight. "Oh, God," she said, "why aren't you getting ready? The only reason I came to Provincetown at all was to go to this party, and you're just sitting there!"

"Oh," said Valentine, "I'm all dressed now."

"This is Garden of Evil night," she reminded him. "So who are you supposed to be?"

"Can't you tell? I'm the Man Who Raped Connie Francis."

Clarisse swallowed the remainder of his drink and ran into her uncle's house.

"I'm glad you brought her down," said Noah. "Clarisse is the only one in the family I still speak to. I'm just surprised you could get her away from that real estate office."

"So was I, in fact. I think she must have been really fed up."

"Why?"

"Low commissions, the owners skimming, the bums making faces at her through the windows—and worse. Besides, she was going to quit in the fall anyway."

"I don't have much to do with family," said Noah after a moment. "Except for her. That's a choice of course, but sometimes you wish you could have things both ways. You know her parents—my brother and his wife?"

14

Valentine didn't answer at once. "I've met them," he said.

Noah nodded agreement to what Valentine did not speak aloud. "And Clarisse's brother is just as bad—worse, in fact, because he's so much younger and he's just the same. Oh well, I—" He broke off suddenly. "Sorry, Daniel, I'm an invariable victim of twilight melancholy."

"Me too," said Valentine. "Listen, I guess I better go in and get ready. Who are you going as?"

"Herod. I'd better get ready too—I think Victor wants to gild my nipples."

An hour later Valentine was again seated by the pool, sipping a third gin and tonic. On the cedar table next to him was a small candle set in a deep glass; the night was black. The candle glow was glossily reflected on his polished brown leather riding boots. His light gray pants were loosely cut; his bloused white linen shirt was collarless and opened to his waist.

He had been amusing himself by watching the White Prince struggle into his Salome costume, when he was disturbed by a noise behind him at the latticed gate. He looked over his shoulder at three middle-aged women who were peering hopefully into the courtyard.

Valentine smiled to himself and rose slowly from the chair. From the cedar table he picked up his coiled bullwhip, jerked suddenly around and cracked the whip violently in the air above the pool. "No!" he shouted. "This is not Poor Richard's Buttery!"

The three women beat a hasty retreat.

Valentine eased back into the chair, but once again there were footsteps on the gravel. He stood poised to repeat the performance, when to his intense surprise, from behind the lattice and hedge emerged a tall, elegantly dressed Oriental woman. She bowed her head slightly, and with the same motion she delicately flicked the tinkling glass wind chimes that were suspended from one of the gateposts. On each of her fingers was a three-inch gold lacquered nail. Her hair was fashioned in a tight chignon and secured at the nape of her neck by large ornamental pins. Her gown was full-length, with red and gold dragons on a field of blue silk. She came a step closer and he saw that she held a small cream-colored envelope.

"I . . ." stammered Valentine.

The Oriental woman broke into sudden laughter. "Fooled you!" she cried. "Fooled you!"

"Damn! How did you get out there without my seeing you?"

"I climbed out the back window," replied Clarisse, coming to his side.

"In *that?*"

"Anything for an entrance. I fooled you!"

"All right, you fooled me. But I've still got to guess who you're supposed to be."

"What do you mean, *guess?*"

"You're . . ." Valentine paused as if in careful thought, and experimentally touched the tape that gave her eyes their slant. "You're Warner Oland—Charlie Chan in drag." Clarisse pushed away his hand. "Maybe not," said Valentine. "Wait. Luise Rainer in *The Good Earth.* That's who it is, isn't it? Except I don't remember that costume."

"That's because Luise Rainer wore sackcloth the whole film, that's why you don't remember this costume, you jerk."

"Who are you then?"

Clarisse flicked her nails at his throat, and then pushed the envelope directly before his face. "Isn't this a clue?"

"Lana Turner in *The Postman Always Rings Twice?*"

"I gave up my job and my apartment to come to Provincetown to live with a cinematic animal," she said in a low voice, shaking her head. "Valentine, I will have you know that I am a near perfect replica of Gale Sondergaard in *The Letter. The Letter,*" she repeated, shaking it in front of his face.

Valentine laughed. "I knew that."

Clarisse sighed. "Light me a cigarette. My hands have been rendered useless for the sake of authenticity." She waved her taloned fingers before him. He gave her one of his Luckys and she seated herself beside him. "Well," she said at last, "who are *you* supposed to be?"

Valentine picked up the bullwhip. "Can't you tell?"

She shook her head.

"I'm Simon Legree."

"You'd make a better Little Eva. All pinafores and sausage curls. Got a date?"

"Name's Terry O'Sullivan."

"Do I know him?"

Valentine shook his head. "He had the other place all last week." Valentine jerked his thumb toward the section of the house behind him.

"How convenient that must have been," remarked Clarisse.

"Convenient for *him*," said Valentine.

"You got chased around the pool?"

Valentine nodded. "Fortunately I was working most of the time. But he was always here waiting when I got off, and always here when I got up, and always knocking on the door to borrow something, or return it, or to ask me if I wanted to go to the beach. I thought, Well there's only a week of it, but then he decided to stay in town a couple of extra days—he took his things over to the Boatslip this morning—just so he could go to the party tonight. When he asked me to go out with him on his last night in town I didn't have the heart to say no. Besides, I'm celebrating the fact that he's getting out of here."

"Who's he going as?"

"Need you ask?"

"Wait a minute," said Clarisse after a moment's reflection, "if you think that I am going to this party with you dressed up as Simon Legree, and some trick of yours in blackface dressed up as Uncle Tom, you are—"

Behind them, there was a loud crunch on the gravel. Clarisse and Valentine turned and saw, kneeling in the open gate, a short man wearing ragged trousers with a rope belt and a soiled yellow vest over a dirty billowing white shirt. His face and hands and bare feet were carefully smeared with burnt cork. His manacled black hands were clasped pleadingly before him, and he said in a loud voice, "Oh, Massa Legree, please don' beat po' Tom no mo'!"

Just at that moment a large party of well-dressed tourists

emerged from Poor Richard's Buttery across the way, and paused to watch Uncle Tom as he advanced, on his knees, across the gravel courtyard. Simon Legree stood and cracked his whip over poor Tom's head, and the Oriental woman, in her great embarrassment, sat with her painted face hidden behind her long-nailed hands.

Chapter 3

Commercial Street at eleven o'clock on Saturday night was even more crowded than it had been at eleven o'clock that morning. The day-trippers had gone home, but their place had been taken by carousing couples in their twenties and drunk teenagers from towns up and down the curved length of Cape Cod. The shops were just closing, and the crowds at the bars just beginning to pick up.

Weaving their way down the crowded sidewalk from Kiley Court, Valentine and Clarisse and Terry O'Sullivan kept a reasonable distance ahead of Noah Lovelace and the White Prince. Noah was splendidly costumed as King Herod in an elaborate mauve robe, its full sleeves and hem bordered in wide bands of silver. His eyes were thickly lined in brown and drawn out in careful spirals at the outside corners. His full mustache was heavily waxed and swept upward; his beard was oiled, and where it was longest beneath his chin, threaded with pearls. A chaste bronze crown rested atop his head and seemed at ease there.

Victor, the White Prince, was done up as Salome. He stumbled along in a pair of spike-heeled open-toed slippers, trying not to trip on the seven voluminous coral veils that were attached to a brown leather girdle riding low on his slender hips. His tanned midriff was bare but for a gaudy green-glass emerald plugging his navel. A halter, also of coral silk, covered his chest, his flesh taped to give the illusion of cleavage, while the cups were filled to considerable capacity

19

with foam falsies. His auburn wig was a masterwork of ribbon curls. On the very top of the White Prince's head was a silver tray on which rested the severed head of John the Baptist—eyes open, mouth agape, ribbons of veins, nerves, flesh, and muscle spilling realistically from the stump of neck. Heads turned, but the Prince ignored startled stares, catcalls, and good-natured derision alike, so intent was he on maintaining his precarious balance.

"Val," said Clarisse, when they were somewhat ahead of Herod and Salome, "why do you suppose Noah puts up with the White Prince? I don't think they've slept together in three years. And sexually, in lifestyles, they've grown so far apart. I just don't understand it."

"Maybe they love each other," suggested Terry O'Sullivan, tugging at his manacles.

Clarisse looked closely at the man in blackface but said nothing.

When Valentine replied, it was to Clarisse. "I don't think Noah's ever thrown anybody out. He probably thinks the White Prince couldn't get along without him. So it's kindness, or habit, or laziness. They still like each other, after all."

"The Prince holds onto Noah like an industrial secret."

"Listen, Clarisse, don't make it your summer project to break those two up. Destroying workable marriages shouldn't be looked on as a pastime."

"I think it's wonderful when lovers stay together for a long time," said Terry O'Sullivan irrepressibly. "I think gay people wouldn't have such a terrible image problem if everybody had a lover, and there wasn't all this sleeping around."

"Yes," replied Valentine blandly, "but on the other hand, promiscuity is a lot of laughs."

"You just say that," said Terry, adjusting his woolly wig, "but you don't really mean it. What happens when you're lonely and depressed? You think some trick is going to get you out of that depression? Everybody ought to have a lover, somebody to come home to, and somebody who helps you wash the dishes, and somebody—"

Here Terry was separated from his companions by a passing knot of lesbians, and Clarisse turned smiling to Valentine. "I think he wants a lover," she remarked. "I think he wants you."

"Poor baby!" sighed Valentine. "We'd have a perfect marriage—for twenty-four hours. But how do I tell him?"

Terry hurried to rejoin them. He was shorter than either. "Have you ever had a lover?" he asked Valentine.

"Yes. But he died. We were *so* much in love. He was all the world to me. Then he was taken hostage in a bank holdup, and the police shot him by mistake."

"Oh, that's terrible! When did all this happen?"

"Two weeks ago," said Clarisse.

Terry O'Sullivan looked from one to the other, dismayed. "Oh, then you must still be very upset," he said to Valentine.

"You want to know how upset he was?" said Clarisse. "He tried to commit suicide with a bottle of NoDoz."

Behind them they heard a shriek, and when they turned it was to see Salome sprawled between two garbage cans in front of a leather shop. Noah was helping him up, and saying in an exasperated voice, "I told you not to wear those damned Spring-o-Lators. They're not even period!"

Chapter 4

Despite the number of tourists that may be found there any time of the year, Provincetown still retains the flavor of a New England fishing village, with tiny cramped houses, sandy yards, narrow streets, and the pervasive smell of the sea—and this is especially true at night, when darkness could make you think you were back in the middle of the nineteenth century. The illusion is dispelled only by the number of roving men on the street after three A.M., and the unsubtle *thump-thump-thump* of rock music from private parties in houses on every other street. Straight tourists think of the town as a trove of quaint architecture, curio shops, restaurants, and guesthouses with never a vacancy glued along the narrow rim of a gray beach. When the sun goes down and the shops close up, these tourists return to the Holiday Inn on the outskirts of town, or drive back to Boston or to other, less expensive towns of the Cape, and have no idea that for many other vacationers, the real excitement of Provincetown is only beginning.

The Garden of Evil party was being held at the Crown dance bar, at the rear of the same compound in which the Throne and Scepter was located. Since it was a private party, it would not be subject to the normal closing time of one A.M., but would probably go on all night.

The Crown was one very large rectangular room with three bars. Overlooking the bay behind the place was a deck, partially covered by an awning that surrounded a swimming pool. Its architecture was utilitarian, and the decoration for the party was straightforward: bowers and festoons of red-and-black paper roses, lanterns that shed flattering red light, and on the walls large blowups of primitive woodcuts, red on black, depicting scenes of rapine, torture, and animalistic butchery. A special slide made up for the bar's laser spelled out in wavering green lines, GARDEN OF EVIL.

The cop on the door, his uniform viewed by many party-goers as yet another costume of evil, nodded them through, but not before giving Clarisse the once-over. She smiled appreciatively. Valentine gave the doorman the invitations he had obtained in his capacity as bartender.

Clarisse touched her friend's cheek affectionately with a sharp nail and headed for the bar. She edged between two sumptuously dressed women and got the bartender's atten-tion by tapping the corner of her letter on his forearm. When she had distributed the drinks, she turned and began with interest to look over the party.

The room was crowded. Costumed figures spilled out and in through the two wide glass doors that led to the deck, edged through into the tiny restrooms, and made an attempt at appearing casual while parading across the lighted stage. Some were immediately identifiable, others were not. There was little dancing yet. With the long nails on her fingers, Clarisse herself was hard put simply to hold a glass.

She turned and looked at the personages on either side of her—the two women dressed in high-forties skirted suits, their hair bound in snoods, with careful tasteful makeup of pearl-shaded powder and ruby lipstick. She furrowed her brow, wondering who they might represent. Then she saw the name tags attached to the lapels of their tailored jackets. HI—I'M EVA BRAUN! and CLARA PETACCI WISHES YOU A PLEASANT EVENING they read. Eva had a period camera around her neck, and begged Clarisse for a photograph. Clarisse graciously linked arms with Mussolini's mistress, and smiled for the Leica's flash.

When Valentine and Terry came up, Eva Braun bent forward and kissed Uncle Tom on the cheek, smudging his makeup. Terry introduced her as Ann, his administrative assistant in a Boston publishing house. In turn, Ann intro-duced her companion, Margaret, from Toronto.

"Oh," said Valentine, "I was looking in your windows earlier this evening."

The two women stared dubiously.

"We're neighbors," Valentine explained. "Clarisse and I have the third portion of Noah's house."

"Oh," said Margaret relieved, "I thought . . ."

Clarisse's brow furrowed. "I think we were all on the ferry together, weren't we?" she asked.

Ann nodded. "Yes. I loved it. They had a great bar. I think bars on boats are great."

"You think bars anywhere are great," snapped Terry.

"Yes, I do. But I especially like bars on boats. The only trouble is you have to drink fast so you don't spill any." As if in illustration, she guzzled off the last of her drink, and set it on the bar to be refilled. Margaret's glass was still quite full.

"We're having a wonderful time here," said Margaret. "We love Provincetown."

Clarisse smiled.

"I want a picture of all of us!" cried Ann enthusiastically. She took the camera from around her neck and asked a man dressed as Charlotte Corday to take a photo of the small group. This done, Terry went to the bathroom to check his blackface and straighten his wig. Eva and Clara left to take a walk around outside on the deck, just as Lizzie Borden swept past, the lace of her bodice speckled with blood, and a small ax tucked into her belt. Valentine followed in her wake. "Oh Val," said Clarisse, grasping his sleeve and pulling him back, "look at Polyphemus over there!"

She pointed to a man standing against the glass wall that looked out onto the deck. His only clothing was a well-fitted loincloth about his narrow waist and a pair of leather sandals laced up to his calves. His shoulders were broad, his chest heavy with muscle beneath a mat of luxuriant tawny hair. Covering his forehead and eyes was a stiff flesh-tone mask, with one stylized eye, glossy and bright, painted in the very center. Slits for his own eyes were effectively concealed in the lines representing folds of flesh. He was talking to Joan Crawford, whose heavy hand rested on the shoulder of a catatonic Christina in white pinafore and yellow sausage curls.

"That is the single most beautiful man I have ever seen in

my entire life," remarked Clarisse. "Dump Uncle Tom, and bring this one home so that I can get a look at him in the morning. I'll pretend that I'm the cleaning lady, and I'll make up the bed while he's still in it, how's that?"

"You missed your chance. I took him home on Wednesday night. His name's Axel Braun—as in brawn."

Clarisse rolled her slanted eyes. "You don't waste any time in this town, do you?"

"A bartender has a reputation to maintain."

"Axel certainly looks like he's worth a repeat performance. Not to mention preservation on videotape."

"He is," said Valentine, exchanging nods with the man in question, "and the nice thing is, his looks are the least of his appeal."

"The *least?*"

"Well, maybe not the least," Valentine admitted. "But in this town—looking like that—he could shoot off arrogance and attitude like fireworks, and get away with it. Last Wednesday night I was down a little, and when I'm down I don't approach *anybody*—much less a man who looks like that. But he came up to *me*."

"Did he need a light?" asked Clarisse. "Directions?"

"It was Janet Gaynor in *Seventh Heaven* all over again. He was very sweet and we had a wonderful time."

"So why is he standing over there, when you're over here?"

"There's a problem."

"What is it?"

"His husband is a jealous woman."

Terry O'Sullivan returned and took Valentine off to be introduced—and shown off—to a group of his publishing friends from Boston. Clarisse ordered another drink and looked over the crowd again, finding it considerably increased. The far end of the room was now a writhing mass of dancers. Clarisse moved slowly and carefully about the room to observe the costumes. Attila the Hun and Anita Bryant made a handsome couple, she dispensing oranges from a large wicker basket on her arm. Charles Manson lingered at the foot of the stage in close conversation with Richard Speck, who had eight blood-spattered nurse's caps dangling from his belt. The Wicked Witch of the West, Patty Hearst as Tanya, the present governor of Massachusetts, Indira Gandhi, Captain Hook, Ramses II, Svengali, Ilse the Beast of Belsen, and Goliath were all there. Clarisse was complimented on her realistic portrayal of Tokyo Rose, Madame Chiang Kai-Shek, and Fu Manchu's daughter.

In passing once, Valentine stopped her and said, "Have a good time tonight, get drunk. You don't start work until ten tomorrow."

Clarisse choked on her scotch. "Tomorrow! Tomorrow is Sunday!"

"This is Provincetown. This is the *season*. Only the dead sleep on Sunday."

"Why didn't you tell me this before?!"

Valentine shrugged. "I was too overcome at seeing you again."

Her glare was ominous, but before she could frame a retort, they were separated by the milling crowds. After a few minutes she found herself not far from Axel Braun. Joan and Christina Crawford had disappeared and a young man, dressed as Ulysses, also in a loincloth and sandals, had taken their place. He carried a sharpened stick. He was small but by no means delicate, with short wavy black hair, and a face that suggested a well-stocked cupboard of resentment and long-treasured grudges. Though she was close to them she could not hear their conversation, for a music speaker was placed just above their heads. Their quick hand movements and rapidly moving mouths, however, implied an argument rather than conversation. This interpretation was confirmed when Ulysses turned sharply on his heel and went out on the deck in the midst of something that Polyphemus was saying to him.

"And that," said Valentine appearing suddenly at Clarisse's shoulder, "was Scott Devoto, the jealous husband. You've just seen round three hundred forty-nine of the longest lovers' battle on record."

"Perhaps we should introduce Polyphemus to *him*," said Clarisse, nodding toward the bar and indicating a young man wearing a white chiton, with a leather sling containing a large stone hanging from his belt.

"I saw Goliath in the men's room. Is that David?"

"No," said Clarisse. "He's got an *X* on his forehead. Must be Cain."

"He's cute."

26

"His name's Jeff King."

"Oh?"

"He's the one from the ferry this afternoon, the one you thought I should have picked up for you."

"I've always trusted your taste in men."

"I wonder if he found somebody to put him up, he was—Oh, Jesus, look!" cried Clarisse, interrupting herself.

She pointed toward the dance area. The music had gone from a heady disco number to a rock piece heavily underscored by an African drumbeat. The center of the area had been cleared and beneath the flashing colored lights, the White Prince as Salome was performing an exotic Dance of the Seven Veils. He whirled and leapt, discarding a veil. He slithered to the floor and writhed, tossing aside another. His mouth gaped, his shoulders undulated, his head lolled—and miraculously, the tray with the severed head was never even jarred.

"It must be sutured to his scalp," said Valentine.

"I'm going to comfort Noah," said Clarisse. Her uncle stood far away and cringed with every round of applause the White Prince received.

Clarisse took Noah out onto the deck, and they settled into sling chairs carefully turned out of sight of the dance floor. They talked of family and the future. The time passed more pleasantly out here than inside, where the decibels and the temperature seemed to increase with every quarter hour.

When word was brought to him that the White Prince wanted to be taken home—not wishing to spoil his hour of triumph by remaining too long in his public's eye— Noah kissed Clarisse good-night and went off. But Clarisse was too comfortable where she was, and just the thought of returning to the crush of the mob inside was enough to bring on the beginning of a headache.

There were a number of persons at the party with whom Clarisse was already acquainted, men and women to whom she had rented apartments in Boston over the past several years, and these she spoke to when they passed near. No one recognized her in the elaborate costume and makeup. Valentine came out periodically and kept her supplied with drinks and lighted cigarettes.

On the deck Clarisse was witness to many matchings—and

almost as many partings. Anita Bryant, very drunk, spilled her oranges into the pool before she passed out in the arms of Atilla the Hun; he eased her into a chair and left the party in the tender embrace of Captain Hook. Patty Hearst accused Indira Gandhi of sneaking around her back with Lizzie Borden, but after some heated discussion Bonnie Parker succeeded in making peace. One who kept reappearing and always in close conversation with someone different was Jeff King as Cain. She'd seen him in one corner speaking low with Ronald Reagan and John Hinckley, and then ten minutes later he was sitting on the edge of the diving board with Cardinal Richelieu. Twice, however, she saw him with Axel —Polyphemus—and their exchange was more than friendly. Axel evidently provided Scott DeVoto ample justification for his jealousy, she reflected. She nodded to Jeff once with a smile, and he smiled politely back, but the puzzle in his luminous cobalt eyes suggested that he did not recognize her.

Toward three in the morning, when the party had thinned to the extent that she did not think her chair would be taken if she left it for a minute, Clarisse stood and went to the railing of the deck. The tide, which had been full an hour before, was beginning to retreat. The black water seemed still and gently reflected the meager lights of the town. Its voice was no louder than the whispers of the couples who walked the narrow beach below. She breathed the cool air deeply and with contentment, actually glad, for the first time that day, that she had come to Provincetown for the summer.

Her reverie was disturbed by a sudden violent scuffling behind her. When she turned around she saw Jeff—his chiton askew—come stumbling out onto the deck, as if he had been pushed from inside. He grabbed one of the posts that supported the awning to break his fall. Scott Devoto stepped out next, his body tense, his face set in anger. He slammed his pointed stick on the deck railing with such force that it broke in two, half of it spiraling off into the air over the beach. Then Axel appeared on the threshold, hands at his sides, still wearing the single-eye mask. None of them took any notice of Clarisse, who had set aside her drink, and stood with folded arms, watching with undisguised curiosity.

"Scott," said Axel sternly, "stop it. You're being stupid."

"You want to go home with *that?*" Scott demanded derisively, looking up and down Jeff's by no means contemptible body. "If you want meat, there's better inside," he said to his lover. "There's that bartender you snuck off with on Wednesday night—go get him. At least *he* didn't give you the clap. If you want chicken, you can find that inside too—and it's younger and cleaner than this piece of shit!"

Jeff looked around uneasily. His eyes locked with Clarisse's, but she did not alter her expression.

"Listen—" began Axel.

"Go on," snarled Scott, in whose voice Clarisse detected a good deal of liquor, "take this one home. But lock the cabinets, and give me your wallet. And tomorrow, go get your shots!"

"The only one I'm taking home is *you,*" cried Polyphemus. He grabbed at the much smaller Ulysses, but Scott dodged him and ran back into the bar. Axel nodded a curt farewell to Jeff.

Jeff watched Axel go back inside, then looked around the deck. He and Clarisse were alone.

"Well," she said, "being left in the lurch is better than a poke in the eye with a pointed stick."

Suddenly he recognized her. "Oh, it's you!" Jeff said, the diffidence and pleasantness with which he had addressed her in the afternoon entirely gone.

"I'm going for a swim," he said in disgust, and with surprising agility he leapt over the deck railing onto the sand, alarming several men who stood directly beneath. Clarisse turned and watched his shadowed figure splash out into the black still water of Provincetown Bay.

She remained at the railing a long moment before she turned away and fumbled to light a cigarette. Before she could accomplish that, Valentine had wandered out to join her.

"How are you doing?" he asked.

"All right. But this party is turning into a parade of bruised bodies and broken hearts. Where's Uncle Tom O'Sullivan?"

"Eva Braun—that's Ann, I think—had too much to drink, and he and Clara Petacci took her out for a little air."

"Have you made your decision yet?"

"What decision?"

"Whether it's to be a single or a double ring ceremony."

Valentine grimaced. "Maybe he'll be satisfied if I just give him the wedding night."

"Oh, dear," sighed Clarisse. "Life is hard when you're an object of universal admiration. Be sure you let Uncle Tom down easy. But in the meantime," she said brightening, "my glass is empty."

Chapter 5

At four thirty, when the disc jockey announced last call, Clarisse was leaning wearily against the deck railing by the pool, an empty glass in one hand and the stub of a half-consumed cigarette in the other. Her chignon had come undone, and her hair fell thickly about her shoulders. The top few buttons of her gown were unfastened, and if she turned to face the breeze out of the east, she felt actually chilled. She closed the eye from which she had lost a contact lens, and scanned the portion of the crowd she could see through the open doors of the bar. She alternately prayed to see Valentine and, in case he had already left with Terry O'Sullivan, cursed him for abandoning her. Despite the DJ's announcement none of the guests seemed disposed to leave the bar; those who had remained this late were determined to bitter-end it, braving the unflattering light of dawn.

Just when she had begun to despair, Valentine and Terry appeared through the glass doors. Terry had his arm possessively around Valentine's waist. Valentine's face and chest were damp with sweat, his linen shirt clung to his arms and back. Most of Terry's black makeup had been wiped off onto his sleeves.

"One for the road?" Val asked.

Clarisse groaned, "I couldn't even see the road."

"Guess you had a good time," remarked Terry pleasantly.

"The blister on my foot came to a head and exploded," said Clarisse. "One of my Bausch and Lomb contacts, which is

31

advertised as eighty percent water, sank into the swimming pool. Someone stole my letter. And after puncturing my wrist three times, I broke every one of the nails on my right hand." She fanned the jagged remains before Terry O'Sullivan's face. "Who's going to call a taxi?"

"We live six blocks away, Clarisse. And the taxis stop running at two."

Clarisse sighed desolately. She set her glass on a table and flicked her cigarette over the deck railing. "Would Uncle Tom like to tote a weary load down Commercial Street? I'll give you a quarter."

A shriek tore across the deck.

"You fucking little bitch!" Joan Crawford roared as she dragged a very drunk Christina to the edge of the pool, stiff-arming aside those in her way. She lifted Christina erect at the edge of the deep end, and then tripped her in. The girl's pinafore billowed whitely about her and the sausage curls formed an aureole of gold around her submerged head. Joan Crawford dropped to her knees, bent far forward over the edge and grabbed a fistful of lank blond hair. She began violently bobbing Christina's head up and down in the water, and pushed away anyone who tried to stop her.

"Life is hard when you're constantly in the public eye," remarked Clarisse. She slipped her arm into Valentine's. "Hold me up. I'm so plastered I couldn't hit the floor with my hat."

It was impossible to get out through the bar. The doorways were jammed with people watching Joan and Christina. At this late hour, however, the bouncer had unlocked the gate that opened to a flight of steps down to the beach, and they escaped that way. As they trudged away across the sand gleeful cheers and the sound of many more bodies splashing into the pool followed them. Terry had moved to the other side of Valentine and once more slipped his arm about the bartender's waist. Clarisse carefully disengaged herself and moved off a few feet.

They walked toward home down the ribbon of hard wet sand left by the retreating tide. "This is probably a lot more pleasant than Commercial Street right now," said Valentine. The houses to their left were black tumbled boxes, with only here and there a lighted window or muffled laughter to indicate that the boxes were inhabited. The early morning breeze lifted Clarisse's hair from her shoulders and dried Valentine and Terry O'Sullivan's glistening faces.

"Isn't this romantic?" said Terry O'Sullivan quietly, and squeezed Valentine. "I feel just like Jeanne Moreau in *Jules and Jim.*"

"Actually that would be *my* role," said Clarisse.

"Oh, look!" cried Terry O'Sullivan, pointing, "it's the morning star!"

"That's a commuter plane," said Valentine. "Did you lose a contact too?"

Clarisse glanced over her shoulder. The sky was lightening behind them. She made them turn and look.

Crossing the municipal parking lot, they nodded friendlily to the fishermen who were already on the way to their boats at the end of the wharf. Terry whispered something to Valentine, and a moment later Valentine in a resigned voice said, "Clarisse, we're going to hurry along. You take your time, and I'll be up before you have to go to work."

Clarisse groaned. "I didn't need to be reminded. You two go on, but remember, Terry," she smiled, "no blood, permanent disfigurement, or toys with combined dimensions of more than thirty-six inches." She waved them on, and the two men took off toward Commercial Street.

Clarisse continued along the beach. She removed her embroidered slippers, lifted the hem of her gown and walked ankle deep in the cool water. The sky was losing its inky blackness, and behind her was a luminous cobalt. Gulls' cries growing sharply louder cut the morning stillness.

Coming upon a mass of seaweed half in the water and half out, she caught sight of a large starfish lying among the thick greenish-blue tendrils. She wondered if she ought to throw it back into the bay, but couldn't remember whether a starfish could live out of the water for any length of time. She also couldn't recall if they stung, or pricked—or just lay there. When she nudged her foot in the seaweed tangled about it and the starfish did not move, the animal—or was it a plant?—took on the character of a souvenir. Clarisse leaned down to examine it more carefully, closing the eye that had lost its contact lens.

It was no starfish, but a human hand.

Then she saw the bare arm beneath the seaweed and the mound she supposed was the rest of the body.

Clarisse stood sharply, and looked all about her. She saw no one. There were only gulls at her back.

She put on her slippers, and walked hurriedly across the sand toward the center of town.

PART II

The Lost and Lonely

Chapter 6

Valentine sat blearily at the kitchen table, one hand wrapped about a blue porcelain mug of strong black coffee and the other resting in his lap. His fingers repeatedly wound and unwound the sash of his green seersucker robe. He lit a cigarette, pulling closer the already butt-laden ashtray. The window beside him was raised several inches admitting the balmy morning breeze from the courtyard beyond. Rampant ivy spilled through, and he regarded the seeking tendrils balefully. Poor Richard's Buttery was serving Sunday brunch, and he could make out the faint tinkling of silverware from the other side of Kiley Court. He glanced at the other two houses in the compound, but all was quiet, and as much as he could make out, no one was stirring. All three flats had attended the Garden of Evil party, and early rising was not expected of anyone. He checked the wall clock above the refrigerator. It was a quarter past nine and Clarisse still had not returned.

He drained his mug and without leaving his chair reached over to the range for the coffee pot and refilled his cup. A crunching of gravel drew his eyes to the courtyard fence. In a moment the ivy on the trellis was shaken and the unlocked gate was jarred violently to the accompaniment of several loud curses before it rasped open. Clarisse stood framed in dappled sunlight through the coffee-tree. Her gown was creased and her makeup had been hastily removed. The silver pins fastened her hair into a ponytail to one side of her neck. She walked across the flagstones in her thin soiled slippers.

Valentine sighed with relief. He rose quickly, filled another mug and set it across the table from him. From the refrigerator he took a walnut coffee ring, retrieved cloth napkins from a cupboard shelf and returned to the table. Clarisse came inside. The screen door slammed against her back and she winced. She kicked off her slippers, pulled the pins from her hair, and performed a little pantomime of plunging them into her heart and plummeting dead into the chair.

They were silent for a minute.

Then Valentine said, "Well, who gets to tell about his night first?"

"Whatever happened to you," she replied, in a hoarse voice and without opening her eyes, "mine was worse. So you go first."

"Well," said Valentine, "when Mr. O'Sullivan and I left you we came directly back here. I was dead. But I thought: we'll have sex, and then we'll go to sleep, and if I'm real lucky, when I wake up he'll be gone."

Clarisse snorted. "Why didn't you just go on and wish for knighthood, undying fame, and the winning lottery number for the next two years?"

"He wanted to talk. He wanted to talk about relationships in general, and ours in particular."

"What relationship?"

"You may well ask that question."

"You should have shoved something in his mouth."

"I did," said Valentine. "But he took it out again. He told me he *knew* that story I had told him about my lover getting killed in a bank holdup wasn't true, and he didn't see why *we* couldn't give it a try."

"What did you do?" said Clarisse, and still with her eyes closed, groped successfully for the coffee cup.

"I said: we either have sex, or we go to sleep, or we say good-bye. He said: I'm not sleepy, and I can't have sex with you until we resolve some of these problems in our relationship."

"Problems? Your first date, and you've got relationship problems?"

"Finally I just gave up, and told him it probably wasn't a

good idea for him to stay—that I didn't think it would work out."

"You wouldn't have had any fun in bed anyway." Clarisse at last opened her eyes.

"He blanched when he saw what I had in the bedside drawer."

He should have looked under the bed," said Clarisse and poked at the coffee cake with a knife. "So after that he left?"

"No," sighed Valentine. "He couldn't believe I was actually asking him to leave. He wanted to talk it all out. I said: 'Go away. Don't come back. Don't call. Cancel your reservation. Move to Canada.' But it didn't get through to him until I actually pushed him out into the courtyard and latched the door. And I hate having to be like that. I'm just glad he's not living *here* anymore."

"He was sweet," said Clarisse mildly. "But I don't think he entirely understood the way you live your life."

"So that was the end of my evening. Tell me what happened to you." He refilled her coffee cup.

She took a long swallow of coffee. "I found a man on the beach last night," she said.

"Good. After a party like that you shouldn't have to go home alone."

"You told your story, let me tell mine. I found a dead man on the beach."

Valentine said nothing.

Clarisse spoke between bites of coffee cake. "I was walking along the beach, and came upon this seaweed, and there was a starfish there—except of course it wasn't a starfish, it was the corpse's hand."

"So what did you do? Did you heave him—*him?*" Clarisse nodded. "—heave him over your shoulder and carry him to the local morgue?"

She looked at him darkly. "I went to get the police. But first I stopped in the ladies' room on the wharf and took off my makeup. Police never take you seriously if you've got on lots of makeup."

"What time was all this?"

"Just after I left you. It probably wasn't even five o'clock."

"It's nine now. What have you been doing for the past four hours?"

"Well, I had to show the police where the body was. And then as long as I was there, I figured I might as well watch. Besides, one of the cops was cute—the same one on the door at the Crown last night. And then they pulled the seaweed off—and lo and behold!"

"What?"

"I knew him."

"What? You mean it was somebody we know?"

"Someone *I* knew. Jeff, surname King. The one I met getting off the ferry. The one who was looking for a place to stay."

"Oh yes. Dressed as Cain last night. In a toga."

"A *chiton,* actually. A toga reaches all the way to the ground."

"A chiton then. What did the police say when it turned out that you knew him?"

"They thought I did it of course," smiled Clarisse proudly. "They said, 'Was this your boyfriend, lady?' For ten minutes I was a prime suspect for Murder One."

"Wait—he was murdered? I thought he just drowned and got washed up on the sand."

Clarisse shook her head. "He was strangled. The police could tell that on the beach. Then they took him back to the station, and made me wait, and then they brought me in to look at him for formal identification. There were bruises on the back of his head—but I couldn't see those very well. He might have been hit over the head with a piece of driftwood or something, or maybe he was thrown in the water and hit his head on a piling. Anyway, there were also purple thumb-marks on his throat."

"Think they'll dust him for fingerprints?"

"They were right against his Adam's Apple. Sort of aubergine."

"What'd you do, compare 'em against a color wheel? How long did you hang around in there?"

"Not long. But I looked close. After all, how often do I get a close-up of a murder victim? Besides, I was still drunk through most of this, and I had to look at him through one eye because I had lost my contact. The police asked me all sorts of questions, but all I knew was what he told me on the pier: that he was in town for the party, and he didn't have a place to stay. I wonder if I'll get in the headlines? *P'town Corpse Identified by Fashionable Woman Attired as Famous Film Star.* Oh God, how can I face reporters with a hangover like this?"

"What about footprints on the beach?"

"Yes, well no doubt the police are going to make plaster casts of the forty-nine thousand prints down there, and they can be almost certain that the killer's is one of them. And then they'll go door to door like Cinderella's prince, making everyone stick their foot in them."

"Wonder what this will do to business?"

"I wouldn't worry. The police were pretty blasé. Once they found out he wasn't heterosexual . . ."

"Figures," said Valentine. "Well, who do you think did it? Who rolled his credits?"

"The last I saw of him, he had just been cast in the role of The Other Woman."

"Who were the co-stars?"

Clarisse smiled and paused for effect. "Polyphemus and Ulysses."

Valentine whistled. "Axel? Scott was up in arms?"

Clarisse described what she had witnessed on the deck of the Crown.

"Did you tell the police about Axel and Scott?" asked Valentine."

"I told them that he had been talking to the men dressed as Polyphemus and Ulysses at the party. I couldn't remember their names. Maybe you ought to call them up, and warn them that the police are out looking for them."

"You think Scott did it?"

Clarisse shrugged. "Let me sleep on it. I'll dream the identity of the killer."

Valentine glanced at the clock. "You have to be at work in half an hour."

"Oh Jesus! I'll have to call in sick."

"On your first day? Beatrice would be very upset. And you haven't even met her yet."

"Call her up. Tell her I have Hepatitis-B. Tell her—tell her the truth. I found a dead body on the beach and I'm reeling with grief because it turned out to be my nearest and dearest friend in all the world, and I've got an appointment with the undertaker to pick out the coffin."

"You can't not show up at work on your first day. Besides, Beatrice wants to explain to you about the merchandise, and then she's going off for the afternoon—you'll *have* to be there."

"But I haven't slept!"

"Neither have I. Saturday night: you spend yours with a corpse, and I spend mine with a man who proposes on the first date."

"I *love* Provincetown." She stood, and began unbuttoning her dress. "Call a taxi and tell him to meet me at the end of the alley in thirty minutes."

"The shop is five minutes away by foot. With Sunday morning crowds, it'd take a taxi twice that long."

Clarisse was in the bathroom. As she closed the door, she shouted, "Star witnesses do not walk to work!"

Chapter 7

Clarisse sat on a high wooden stool behind one of the four glass display cases that were arranged fortresslike in the center of the Provincetown Crafts Boutique. Her hair was arranged in an efficient bun; she wore a white silk blouse, dark brown waist-pleated slacks, and sensible low-heeled shoes. It was an outfit she'd thought appropriate for appearance behind the counter of a shop specializing in "rare and beautiful things"—Valentine's words. Her folded arms rested heavily against the edge of the beveled surface; her clouded eyes shifted uncertainly and unhappily about the room. On the back of a receipt book she quickly made a list of four terrible ways for Daniel Valentine to die.

Of approximately two thousand objects offered for sale and individually priced, not one was either rare or beautiful. On tables crowded against the inside wall, dozens of gaudily painted porcelain clowns with leering smiles faced an army of machine-carved Cape Cod fishermen brandishing vibrant red lobsters. Plaster fish were stacked in four-foot pyramids on either side of the door, and in a water-filled bucket just in front of the cash register a blue plastic whale endlessly swam around and around and now and then spurted a geyser of water through a blowhole in its head. On the wall were posters celebrating the glory of summer on Cape Cod, and printed Chinese calendars on bamboo scrolls. On the door and window frames were tiny mirrors in seashell-littered frames. Suspended in the large many-paned front window was a profusion of stained-glass plaques of sailboats, sunsets, and endangered species. The ceiling was a tinkling sea of glass, bamboo, and metal wind chimes. Everything had "Olde Cape Cod" stamped prominently on its bottom or back. Clarisse began a game in which she searched the shop for a color that was found in nature, but soon gave it up as a waste of time.

She had been so stunned when she first saw the shop that she had been scarcely able to reply to Beatrice Rowell's pleasantries when Valentine had introduced the two women. Beatrice, the shop's owner, was probably forty-five, and something about her *screamed* divorcée, but she was very pleasant, and dressed—if not with flair—then at least not without taste. Clarisse thought perhaps that the Provincetown Crafts Boutique was some sort of elaborate joke, with Beatrice laughing every time a tourist made a purchase. Clarisse wasn't laughing yet.

She picked up the instruction booklet for the machine in the corner which heat-pressed decals onto T-shirts, but then reflected that perhaps she ought first to master the intricacies of the electronic cash register, which after half an hour's experimentation, she hadn't yet been able to open.

It was nearly eleven o'clock and the tourists were just beginning to show themselves on the street in oppressive numbers. Each time someone entered the shop, a music-box chime played the first four notes of "Lara's Theme." Clarisse experimented with keeping the door propped open, but the early heat of the day was unpleasant and she satisfied herself with disconnecting the wire that triggered the music. When a customer wanted to know the price of one of the carved fishermen—which was clearly marked $2.98—Clarisse replied, "It's forty-five dollars, not including tax," because she was afraid the man would want to buy it and she would have to admit her ignorance of the cash register.

The door of the shop was held open for this rapidly retreating customer by a policeman. He turned to a fellow officer who was standing outside and said something which Clarisse did not hear. The fellow officer wandered off, and the handsome policeman—Clarisse had rapidly determined *that*—came inside with a smile.

He stepped up to the counter, rested his hands flat on the edge, and let his eyes sweep boldly over her. Since Clarisse had already done the same for him, she allowed her eyes to remain demurely on his face. He was tall and slender with skin that looked incapable of burning or blemishing. His closely cut wavy black hair framed a strongly featured face with high cheekbones, a sensual pouty mouth, and large nearly black eyes bordered with heavy lashes. His uniform, unlike those of the other cops she'd seen in town, seemed tailored to fit his body.

"The last time I saw you," said Clarisse, "there was a corpse between us."

He nodded, but said nothing.

"Have you come to take me away from all this?" She glanced around the shop with a little grimace. "For questioning, I mean?"

"My name is Matteo Montalvo." He spoke his own name with an accent, and Clarisse immediately conjectured that he had been raised in Provincetown's Portuguese community. "Call me Matt though."

"I like Matteo better," said Clarisse. "Do you want me to lock up?" she asked. "And come down to the station with you for a few hours? Say until closing time, whenever that is?"

"No," he smiled, and glanced around the shop. "I wish I worked in a place as nice as this." Clarisse charitably supposed that he referred to the air conditioning. "I came to ask you out."

Clarisse paused only a moment before answering. "Sure! Just let me close up." She hopped down off the stool.

"No!" laughed Matteo, "Wednesday night."

Clarisse sighed, then looked up. "Oh," she sighed disappointed, "and I thought it was my first invitation to breakfast with a cop. . . ."

Just before noon, when his shift began at the Throne and Scepter across the way from the Provincetown Crafts Boutique, Valentine came into the shop, bringing a tongue sandwich and a bottle of Saratoga Water for her. "Thank you for the sandwich," Clarisse said with a lowering glance which Valentine ignored. "And for the position. It's a shopgirl's dream."

"As soon as I saw what Beatrice was stocking, I knew it was for you."

A woman customer pointed to a small jewelry box completely covered with tiny scallop shells. "How much?"

Clarisse consulted a printed list of prices taped to the counter. "Ninety-four-fifty."

The woman grabbed her husband's sleeve and pulled him out of the shop.

"Nothing in this place is over fifteen dollars," said Valentine. "Let me see that price list."

Clarisse explained her problems with the register and Valentine, with little difficulty, showed her how it worked. "Well that wasn't hard," she said. "I thought you had to have a degree in higher math for these things. Come back on your break and show me how to operate the T-shirt press."

He leaned on the counter and began to eat a turkey club that he had brought for himself. He occasionally helped himself to a drink from Clarisse's bottle of Saratoga Water.

"So," said Clarisse, "is the whole town buzzing with news of the murder? Have people been asking if I've recovered from the trauma yet?"

"Well," said Valentine, "I ran into a couple of people on my way here, and I talked to George while he was making the sandwiches. At least everybody seems to have heard. Nobody mentioned you though."

"I guess it takes time for these things to get around. By the end of the day this place will be mobbed with people asking me what it felt like to discover a corpse at sunrise. I'll be the heroine of the hour."

"Maybe," said Valentine doubtfully.

"I just wonder if we'll ever find out anything about Jeff King and what he was doing yesterday—in the last hours of his life I mean."

Valentine shrugged. "I heard a little about *that*. . . ."

"Oh?"

"Well, after he left you on the pier, fresh off the ferry from Boston, he went to the Boatslip. In the ladies' room he changed into a pair of green swim briefs. They were very low-cut in the back. He swam in the pool for about half an hour but didn't do any diving. He had two drinks—both vodka and tonics. And he was selling drugs by the poolside out of his gym bag. MDA for sixty a gram, Black Beauties for seven a tab, and crystal coke for one-twenty-five a gram. I couldn't find out how many contacts he made. He went to Ciro and Sal's for dinner. Jimmy waited on him. He had eggplant parmesan and a glass of wine, house dressing on his salad—not bad for a last meal. He had dessert at the Portuguese bakery across from the candy store. Nobody knows for sure what he bought, but it was probably a cannoli with almond filling. He went back to the Boatslip, changed into his costume—but not in the ladies' room—and went to the party. He paid the cover charge with a hundred-dollar bill but didn't leave any tips when he got drinks."

Clarisse paused a moment. "Did he leave a printed itinerary?"

"People are talking. News got out before most people had gone to bed."

"I just don't understand," moaned Clarisse. "There's a minute-by-minute account of the corpse's doings and goings, and nobody mentioned *me*? I mean, I *found* him. I even *touched* him. Maybe I'll get in the papers."

"I don't think so," said Valentine. "Murder is bad for business. Everything'll be handled very quietly."

"Well, the police came to see me again a little while ago," said Clarisse.

"I thought they got your statement this morning. You were at the station long enough to write your autobiography."

"No, the cop had only one question to ask."

"What was that?"

"'Do you want to go out Wednesday night?' Wednesday night's his night off."

"He came to ask you for a date? I never thought a woman appeared to advantage when she was identifying a corpse."

"He saw through the superficialities of the situation. Anyway, I said yes."

Valentine's eyes widened. "Was he cute?"

"I could spend the rest of my life with that man," sighed Clarisse. "And he was Portuguese too—you know what *that* means."

"Yeah," said Valentine. "All Portuguese men have big—"

"Hearts," said Clarisse quickly.

"But this cop didn't mention the murder?"

"No," said Clarisse. "I wonder if they even know that Mr. King was selling drugs. I wonder if they know any of what you told me."

"Probably not. The entire FBI doesn't have the investigative powers of five gossiping queens."

"There's something you said that I don't understand: How do you know Jeff King didn't change into his costume in the ladies' room at the Boatslip?"

"Because there was a big contingent of dykes there late yesterday afternoon, and it was off-limits to men. My friend Larry was sitting on the porch of the Casablanca all afternoon —that's right across the street—and he saw Jeff go in with his bag, and come out later in his costume, but without the bag. So somewhere in the Boatslip he changed his clothes."

"That means he was probably staying in somebody's room. Maybe he tricked," suggested Clarisse.

"If it was only a trick he wouldn't have left his bag. It was more likely he ran into somebody who agreed to put him up. Somebody he already knew. That also means that his bag is still somewhere in the Boatslip."

"So at least there's *one* person who's not talking," said Clarisse ruefully. "I just wish all these people would get the really important part right."

Valentine looked up from his sandwich curiously.

"The really important part," said Clarisse, "is that *I* found him, alerted the police, and provided positive identification."

"Give up, doll," said Valentine. "This dead drug dealer isn't going to make you famous. I wasn't going to tell you this, but the rumor is that the body was found by an antique dealer from Chicago who was on a bad acid trip, and it sent him over the edge and he had to be flown to the hospital in Hyannis."

Clarissa was incensed. "Oh," she cried, "stealing my thunder! Don't talk to me about the detective abilities of the male homosexual. My one chance to get my picture on the cover of *Real Detective* and *True Police Stories*—and the gay community can't get it straight!"

Chapter 8

Shortly after Valentine had left the shop, Beatrice returned to look in on Clarisse.

She opened the door, stuck in her head, pulled her dark glasses down low on the bridge of her nose, and peered at Clarisse over the top of them. "How are you getting along?"

"Just fine," replied Clarisse with a grin that was as wide as it was insincere.

Beatrice came all the way into the shop. She wore a forest-green linen dress with black trim and a black sash. Her sandals were lacquered black. Her skin was of a leathered toughness that is attained only in women of a certain age who have spent nearly all of their lives in Florida or southern California.

"I'm so glad you've come to work for me," said Beatrice. "I'm so happy that Danny found you for me."

"Danny?"

"Dan Valentine," said Beatrice with surprise.

"On, *Danny*," replied Clarisse. "Yes, well, I needed the job."

"You're from Boston?"

Clarisse nodded.

"Did you work in retail there?"

"I was the principal legal consultant for a major advertising firm with offices in Boston, San Juan, and Honolulu."

"And you gave that up to come work *here?*"

"I was tired of the grind," said Clarisse with a confiding nod. "The constant travel to Puerto Rico and Hawaii wore me out."

"I'm afraid you'll find it very dull in the Provincetown Crafts Boutique then," said Beatrice, with some concern in her voice. "But I hope you find it exciting enough to stay on for the whole summer. I had a girl in here, hired her on April Fools' Day—and I was the fool. She robbed me blind."

"She knew how to work the register, then?"

"No," said Beatrice with widened eyes, "she was stealing the *merchandise*."

Clarisse looked around. "Are you sure?"

"I caught her one day pushing a ceramic toothbrush holder down the front of her pants. Then I went to see her parents, and they showed me her room. It was *filled* with things from this shop. Her parents were very upset. Judy had told them she got them as commissions for doing so much selling."

"A wicked girl!"

"I brought everything back," said Beatrice, "and put it on a markdown table. And do you know that that girl had the effrontery to come back in here and *buy* the pieces that she liked best?"

"What she did was very wrong," said Clarisse.

Customers entered. Beatrice smiled at them, and stepped closer to the counter. She said to Clarisse, "I'm flying up to Boston this afternoon to go to the gift show at Hynes Auditorium." She sighed happily. "The gift show is like Aladdin's Cave to me. Can you imagine this place expanded to the size of twenty-five basketball courts?"

"No," said Clarisse quickly. "I can't."

"*That's* what the gift show is like. I should be back by seven and I'll bring you some catalogs to look through. You can help me choose some new stock."

"I'd love to," said Clarisse, with a smile that was genuine.

Clarisse liked Beatrice, though she could by no means endorse the woman's taste in bric-a-brac. Clarisse did her best to wait cheerfully on customers that afternoon and not condemn them as Philistines simply because they wandered in to browse. But her resolution wavered as the crowds grew larger and the hour grew later. When Beatrice returned at the promised time and took over for the rest of the evening, Clarisse staggered across to the Throne and Scepter, where Valentine still had an hour on duty.

"This was the worst day I've spent since I heard that Patty P. Hearst had been kidnapped," she whispered, and groped blindly for the drink he'd prepared for her.

Valentine smiled. "Too bad you had to begin like this, but Sunday's always the busiest day around here. Probably during the week everything'll be very quiet and pleasant—"

"The Provincetown Crafts Boutique could *never* be 'quiet and pleasant.' Not with *that* merchandise. I felt as if I were presiding over an elves' workshop in there."

"Well, just relax. After I finish here we're going home and change clothes and I'm taking you out to dinner. I even"—he opened the refrigerator behind the bar and took out a small box and held it up to her—"bought you a corsage. Symbidium."

Clarisse shook her head slowly. "As bribes go, it falls short of a proposal of marriage or a shoeboxful of diamonds, but I suppose it will have to do."

The Throne and Scepter was not crowded. Many of the visitors who had come to Provincetown for the day, the weekend, or the previous week were packing up now or had already left; the town seemed quiet. At a table just behind where Clarisse sat at the bar two men in their late forties were fighting, merely for the pleasure of it, it seemed. Their relationship had broken up formally eight years before but they still debated the causes and the blame, and seemed very pleased that Clarisse was attending closely to them.

To her right at the bar were seven men, of greatly varying age and appearance but all Provincetown regulars, engaged in a kind of round-table discussion on who had the biggest tits in Hollywood. The contest had narrowed to Kathryn Grayson and Mamie Van Doren, with Miss Jane Russell contemptuously dismissed as publicity hype. When Valentine placed a second drink before Clarisse, one of the men turned to her and asked, "You ever had a screen test, honey?"

When she went to the ladies' room, Clarisse passed through a dark corner of the bar, and to her surprise, discovered Ann and Margaret sitting at a tiny table that was nearly hid behind a vast palm in an Art Nouveau pot. Holding hands and gazing intently into each other's eyes, they did not even notice her until she spoke.

"Good afternoon," said Clarisse pleasantly.

The two women looked up, grasped for recognition, and then broke into smiles.

"Hi," said Ann, lifting her drink in a toast.

"Hello, Clarisse," said Margaret with a smile.

They spoke for a moment about the party, about Noah's pool, about their plans for the evening, then Clarisse went on into the ladies' room. When she came out again, Ann said, "Have Daniel bring me another gin and tonic, will you please? He's been forgetting me."

"No, he hasn't," said Margaret in a low voice. "You've had enough. If you have any more, you're not going to be able to *taste* your dinner."

"I want one more!" protested Ann.

Margaret sighed and nodded to Clarisse. "Have Daniel send one over, and a Perrier for me."

Clarisse walked away, and heard the two women buzzing behind her. When she got to the bar, she said, "The lady who doesn't need another gin and tonic wants another gin and tonic."

"That's six," said Valentine, shaking his head. "Do you think she'll try to bust up the place?"

"I like to drink," said Clarisse. "But I think it's undignified for a woman with an appearance to maintain to fall on her face before eight o'clock. Barroom floors always smudge your makeup."

"If I had Terry O'Sullivan for a boss, I'd get sloshed on my vacation too."

"Margaret is trying to keep her in line, that's something. Is this summer love or is it a real affair, do you think?"

"Summer love. Unfortunately, they're involved in a four-sided triangle." Clarisse turned to him inquiringly. "Ann will have to go back to Miriam in Boston. Miriam has a lot of money and an ugly temper."

"And Margaret?"

"Is married to Joyce, in Toronto. Joyce is thin, and supports her mother in a nursing home."

"How do you know all this?"

Valentine shrugged. "A good bartender learns a little something about every one of his customers."

"You eavesdrop, you mean."

Valentine held up his hands in protest. "I draw the line at mechanical listening devices. I scorn hidden microphones. All my information is obtained legally. This information came from Mr. Terry O'Sullivan." He moved away to wait on Mamie Van Doren's most fervent partisan. In the dark corner of the bar, Ann burst into tears and fled into the ladies' room.

Margaret came to the bar. "Tell Daniel to forget about the reorder, Clarisse. Ann and I are leaving. You don't happen to know where I could pick up a home Detox Unit, do you?"

Chapter 9

A little later Valentine and Clarisse were walking back up Commercial Street from Kiley Court. Valentine wore a loose-fitting white summer suit circa 1940 with a black shirt printed with a single line of enormous long-stemmed yellow roses. Clarisse wore a white dress of the same period with the spray of symbidium pinned to her bodice. She'd fashioned her hair into a style in imitation of one worn by Eva Perón. Their appearance as a sterling couple of fashion and consequence was undermined only by Valentine's winking at every good-looking man that passed.

After the madness of Saturday night and Sunday afternoon, the streets seemed almost deserted. The day's blasting heat had abated beneath a balmy salt breeze that wafted across Commercial Street from the bay.

The Swiss Miss in Exile was a small two-story Victorian house, set well back from the street which had been renovated into a fair likeness of a Swiss chalet, with pierced shutters and a great deal of gingerbread. It was painted raw sienna and canary yellow, and its window boxes were filled with red geraniums. Daniel led Clarisse up the evergreen-lined path toward the entrance.

She paused at the threshold and glanced at a couple of grinning stone dwarfs that stood bowing at either side of the door. "I've never eaten here before," she remarked meaningfully. "Swear to God that the food will make up for the decor?"

"Food's good," said Valentine, stepping into the front parlor. In this room was the maitre d's desk, the reservation book open on it, and several comfortable chairs for guests waiting to be seated. "But don't you know why I brought you here?"

"You're meeting a boyfriend who's into dirndls?"

Valentine shook his head, and lit cigarettes for them. The maitre d' hadn't yet appeared. "Your uncle *owns* this restaurant."

"What!"

"He bought it last January, and then had it fixed up. I forget what it was before—a guesthouse I think. It wasn't gay so of course it went under."

"You mean to tell me that Noah *authorized* those charming architectural details on the facade of this building?"

Valentine pointed to the bright red-and-green stenciled walls in the reception room: "And the interior decoration as well."

"*Why?* Noah keeps his business dealings pretty much secret, but I didn't think he knew anything about restaurants —or does he?"

Valentine leaned forward and whispered, "Maybe not, but the White Prince does. . . ."

Clarisse nodded with sudden understanding. "And *that's* why he's never mentioned it to me, I'll bet. So Noah invested let's say fifty thousand dollars to keep the White Prince happy. I might have known. Why doesn't Noah want to make *me* happy? For only twenty-five dollars he could buy me a sledgehammer for the Provincetown Crafts Boutique." She looked around her with increased interest. "It's probably doing all right, too. Noah's never lost money at anything he did."

"And the White Prince has never made any," Valentine reminded her.

"God. At least he's not the maitre d'. If he were, straight customers would never get seated."

"I think he's mostly kept out of sight. Even though he looks as though the only pencil he ever used was to do his eyebrows, the White Prince is actually pretty good with books."

Beyond the front parlor, in the warren of large and small rooms on the first and second floors, from two to seven tables had been set up in each, and in the backyard, made private by an old and vigorously pruned privet hedge, there was garden dining. Presently, the maitre d', wearing a saffron-hued peasant shirt and raw-cotton slacks, seated them in a tiny room overlooking the garden. The breeze through the lace-curtained window was warm and fragrant. Their waiter was tall and sufficiently handsome, clad in a white shirt, knee socks, clogs, and chocolate shorts held up by brightly embroidered suspenders. They settled comfortably into high-backed rush chairs, noted with satisfaction that the only other table in the room was unoccupied, and ordered drinks. The illumination was provided by candles only—on the mantel behind Daniel, in sconces behind Clarisse, and in a yellow glass globe on the table between them.

"Thank you," said Clarisse to Valentine when the drinks were brought. She raised her glass. "This is just what I needed after today."

"I was thinking about taking you to the Forward Pass, but I wasn't sure you'd be up for waiters dressed like cheerleaders."

"No," she said thoughtfully, "probably not." She placed her clutch bag on the table, and cautiously lifted the lid of a small box next to the saltcellar. It played a tinny Viennese waltz. She slammed the lid shut. "Candlelight," she said. "And a large menu, and a waiter who knows what he's doing—that's what I needed, having been so recently subjected to the brutal side of human nature."

"Your customers weren't that bad."

"I'm talking about Jeff King."

"Are you going into your *Witness for the Prosecution* routine again?"

The waiter returned. Clarisse said, "Order for me, Val. I'm in no condition to make minor decisions."

Valentine spoke to the waiter for a few moments, and when he was gone, leaned forward and pulled back the lace curtain from the window. The last moments of twilight hovered over the garden. Yellow lamps placed in niches carved in the privet hedge lighted the area softly. The murmur of conversation and the discreet clatter of dishes and cutlery was very pleasant.

"There's a cutie," said Valentine, and pointed to a man seated alone at a table in the corner of the garden.

Clarisse peered out. "How can you tell? He's got his back to us. And he's in almost total shadow."

"Sea air sharpens my senses. I can *smell* a cutie—especially when he's got shoulders like that."

"Maybe if I smashed a window he'd turn around and you could get a look at his face."

"I know those shoulders, in fact," said Valentine.

"You would. Who is it?"

Valentine paused for a moment, considering. "It's Axel Braun," he said.

"At the party? Polyphemus?"

"And where's Ulysses I wonder," mused Daniel.

"I don't know," said Clarisse, "but I'll bet you he's not out laying flowers on Jeff King's grave." She peered out the window again. "Axel looks depressed."

"He's got his back to us. How can you tell he's depressed?"

"All good-looking men get depressed on Sunday night, especially if they're alone. I know it for a fact."

At that moment, Axel Braun, holding a glass of wine, turned in his chair and looked toward the doorway to the interior of the restaurant, as if hoping to see someone there. He turned back after a moment, slightly hunching his recognizable shoulders as he did so.

Through appetizer, salad, and entree, Clarisse caught Valentine up on Boston gossip, detailed her plans to attend the Portia School of Law in the autumn, and then confided her intention of destroying at least one item a day in the Provincetown Crafts Boutique. "The clowns are easy because they're plaster. You just sort of push one off on your way to the storeroom. It could happen to anybody. When they're all gone, I'll start on the fishermen, but they're a lot harder, because they're made out of wood. But what I'd *really* like to get at are the wind chimes, but that's almost impossible to do 'accidentally,' because I have to stand on a chair to reach them."

When their dishes were being cleared away, and they were finishing off the Burgundy, Clarisse said, "Oh, I meant to ask you, did you hear anything more about Jeff King? Did you fill in the gaps of his itinerary?"

"People did talk," replied Valentine. "But by the after-noon the rumors were starting to get wild. I wonder if we ought to believe all the things we heard this morning."

"Probably not. But still—tell me what you heard at the bar."

"Well, that Jeff King not only had the bridal suite—there isn't one—at the Boatslip, that he had a room at the Casablanca, the Pilgrim House, and the Holiday Inn at Truro. He told five people he thought someone was trying to kill him, had sex with seventeen men in the dunes, danced with at least six dozen people at the Boatslip and at the party later, dealt about one ton of drugs including angel dust and heroin, got drunk, threw up, hallucinated himself into a frenzy, and *still* managed to get himself killed on the beach—all in about sixteen hours. He OD'd on heroin, had a live grenade shoved in his mouth, he was spiked to death by a drag. But of all the people I talked to, no one seemed very sorry—it was odd."

"Why, do you suppose?"

"Because Jeff King had the bad sense to get murdered in a resort. Nobody knew him, nobody knew who he was. If these people were back in Boston or New Haven or wherever and a gay man had been killed a few blocks away, they'd be up in arms. But down here they figure . . ." Val shrugged.

"Figure what?" demanded Clarisse.

"They figure: Who cares? They come here and they say: Entertain me. Give me a good time. I'm paying enough for it. Well, entertainment doesn't include worrying yourself sick over a drug dealer who turned the tide red on Sunday morning."

Clarisse was silent a moment. Then she said, "That's a rotten attitude."

" 'Who cares?' " repeated Valentine softly.

Chapter 10

"Yooooodeeelooooowwwdeeelaaayyyyheeehooooo!!!" rippled majestically from outside the door of the dining room where Valentine and Clarisse sat waiting for the dessert menu.

"That wasn't recorded," said Clarisse darkly, and turned in her chair a little.

A face like a full glimmering summer moon popped into view at the side of the doorframe. Below it Clarisse could see a great flounce of periwinkle blue material supported by stiff white organdy underslips. At the threshold peeked a round white-stockinged foot in a pink satin slipper.

"Quick, Val," Clarisse murmured, "pour me the rest of that wine."

With a second yodel that was longer than the first and by far louder, the woman backed into the room, hunched over and with mincing little steps, grinning at Valentine and Clarisse over her shoulder. Her face was round and pretty, with wide bright green eyes beneath thick fluttering lashes. Her mouth was bright red and bee-stung, the globes of her cheeks rouged into soft circles of rose. Her hair was honey blond, parted in the center and twined into braids that dangled like mooring ropes below her waist.

She turned around and cried, "Yah, goot evenink!" She beamed, picking at the puffed sleeves of her white blouse and yanking at the straps of her pinafore.

59

Valentine had turned toward the window, leaning an elbow on the table and biting his knuckles to keep from laughing.

"Good evening," said Clarisse weakly.

"Yah, yah!" the woman cried robustly. "You vould like to see de lingunberry tart oder de *mousse au chocolat* oder the pfeffernuesse oder . . . ." When she moved several steps to the side, Clarisse at last could see the dessert cart that had been hidden behind her. She threw her braids over her shoulders, nearly knocking the breath out of a waiter who was passing through the room, and pulled the cart up to the table. "Oder vielleicht," she went on, "you would like de Heidi Sviss Miss Special?" She pointed to the centerpiece, a small Mont Blanc with the minuscule figure of a little girl—dressed exactly like the woman herself—skiing down the side of the confectionary mountain.

The woman fluttered her eyes madly at Clarisse and shifted her weight from one foot to the other. The cart rolled just a little when she did. "Take de time, yah!"

"I'm not certain that I'm ready for this," said Clarisse softly, looking at Valentine and pressing his foot beneath the table.

"Vat?" cried the woman, "not even yust von?"

Clarisse shook her head hesitantly. Valentine broke into laughter, stood and kissed the woman on her cheek. "Clarisse," he said, "this is Angel Smith."

Angel Smith grabbed the hand that Clarisse was about to offer to her and shook it frantically. "Nice to meet you," she said, without a trace of accent.

She pushed the cart out of the way, pulled up a couple of chairs and seated herself at their table. "I'll confess," she said, glancing back at the dessert cart, "the pudding is from yesterday and it's a little rubbery, but the strawberry tarts are great. I whip the cream myself." She hoisted three of the large tarts onto the table, grabbed three forks and three spoons from a tray beneath, and while Clarisse and Valentine murmured their thanks she began taking birdlike dips of the whipped cream. "Heaven," she breathed, "just heaven."

"How have you been?" asked Valentine politely.

Angel groaned between bites. "Business is great, but I'd like to bury Heidi." She looked at Clarisse. "Do you know how humiliating it can be for a thirty-two-year-old woman to dress up like this every night just to flog desserts? With whipped cream like this people would buy them even if it was Nancy Reagan pushing the cart. Oh, well," she conceded, "got to make a living." She popped three strawberries into her mouth.

"Do you work in Provincetown every summer?" Clarisse asked.

"Like cuckoo-clockwork."

"Perhaps you could find a job in a restaurant where you didn't have to dress in a ludicrous costume."

"Ah—Clarisse . . ." Valentine cautioned.

"I think of what my mother would say," said Angel, putting down her fork. "But the fact is, I *am* the Swiss Miss. And when I'm in P'town, I'm the Swiss Miss in Exile."

Clarisse shook her head. "I don't understand."

"Angel and Noah are partners," said Valentine.

"See, the Swiss Miss in Exile is an offshoot of our first Swiss Miss, on Harvard Avenue in Brookline next to that all-night bagel palace."

"Where is the White Prince?" asked Valentine. "I haven't seen him framed in any doorways this evening."

"One of our dishwashers ran off with a fabulously wealthy nature photographer he met in the dunes, so the Prince is downstairs in the kitchen screaming like Fay Wray."

"I was hoping to run into you at the party last night," said Valentine.

"Oh, yes! How was the party? I was dying to go—" She shifted her eyes to Clarisse. "I was going as Pat Nixon. Catch the resemblance? I have this thing about First Ladies." She went back to Valentine. "But I didn't get into town until this morning, so I missed all the fun." As she scraped the last of her tart free of the dish, she glanced speculatively back at the dessert cart. "I heard there was some trouble," she said offhandedly. "Someone got killed?"

"That's all you heard?" asked Clarisse. "I thought it was all anybody was talking about."

"I've had my head in the oven all day."

Valentine provided Angel with a brief account of the death of the man whom Clarisse had met getting off the ferry.

"So," said Angel, turning to Clarisse at the last, "*you* found him! That calls for another dessert!" She hauled the

Mont Blanc from the cart and placed it before Clarisse, who hadn't managed even to finish her strawberry tart.

Clarisse smiled and pushed the Mont Blanc toward Angel, who shrugged and dug in. "Did you try to revive him?" she asked after half a dozen quick bites.

"It was too late for that," said Clarisse.

"Are you sure? Honey, if it had been me finding a man alone on the beach at that hour, I would at least have tried a little mouth-to-crotch resuscitation."

"Mr. King was cute," said Clarisse, "but he was a goner."

Angel put down her spoon. "His name was King? And his first name was Jeff?"

"You knew him?" asked Valentine.

"Once upon a time," said Angel, plucking Heidi from off the mountain, "I knew a Jeff King. Describe the corpse, please," she said to Clarisse.

"Clone. Cute, but still a clone. Short dark hair, well-trimmed mustache, standard-issue body. Except for his eyes —his eyes were different. They were cobalt blue."

Angel shook her head ruefully, but didn't put down her spoon. "Same one."

"You *used* to know him," prompted Valentine.

"Hadn't seen him in seven years. Not since I lived on Queensbury Street. That was a bad time for me," she said seriously. "I had no money, no prospects, and a boyfriend whose favorite colors were black and blue. I'd go down to the Haymarket late every Saturday afternoon when they were throwing away all the stuff that they hadn't sold—and that was my weekly shopping. It was," she said delicately, "the worst two years of my life."

Valentine and Clarisse were silent.

"Anyway," Angel went on, "right under me"—she paused a moment, then began again—"this nice man lived in the apartment right below mine. Gay, handsome, about forty— but he had this kid living with him. And that kid was Jeff King. He didn't look like a clone then, of course—he had long hair and a beard, your basic student hippie. Also your basic thief. I'd come in in the afternoon—I was working part-time at the Burger King around the corner—and I'd find him hanging around the mailboxes. One month this old lady upstairs didn't get her Social Security check. I had to steal from Burger King for the entire month just to keep her alive. And then the next month, my welfare check didn't come. I reported it, of course, but they didn't believe me, not even when I showed them that the signature on the canceled check wasn't mine."

"And you're saying it was Jeff King who stole those checks?" asked Clarisse. "How did you know for sure?"

"I just *knew*. After that I'd always glare at him in the hall, but he'd never look at me. Sometimes after that he skipped out on the man downstairs, stole an emerald ring and eight place settings of Rosenthal china. All that was seven years ago, and I'm doing fine right now, but I'll tell you something —I'm not sorry he got his. In fact, I wish it had been *me* who found him. Do you know what it's like to go without money for a month, I mean to have *no* money for an entire month? And all the little bastard wanted that money for was to buy drugs! He stole my welfare check and bought a gram and a half of cocaine! I hope he suffered when he died." She plunged her fork into the middle of Mont Blanc, splitting it open like an earthquake.

Valentine said nothing. Clarisse fumbled a cigarette out of her bag and lighted it.

Angel heaved a great sigh and erased the consternation from her face. "Oh, now I've ruined your evening," she sighed.

"No," said Clarisse softly. "Not at all."

Angel pushed back her chairs and stood. She waved a hand over the table. "This evening's on the house. No back talk, Daniel, or I'll clog-dance right here—in double time, with wings and everything." She turned to Clarisse, and her face was suddenly stern once more. "I feel like an old score has finally been settled. So tonight I'm celebrating."

Chapter 11

Valentine and Clarisse were about to get up from the table, when their waiter suddenly appeared with liqueurs, compliments of their hostess, whose yodel they heard in the garden. Clarisse pulled back the curtain and smiled at Angel Smith, who grinned and waved. She dropped the curtain into place, then said in a low voice to Daniel, "Here comes the lost and lonely. Don't let him get by without finding out what he knows."

Axel Braun had entered the house from the garden and would have left without seeing them had not Valentine quickly stood and fetched him from the adjoining room. "Come on in and help us kill this bottle of Drambuie," he said, and drew Axel into the tiny dining room. Clarisse signaled the waiter for the third glass, and smiled at Axel.

Valentine introduced them. "I'm Val's roomie," said Clarisse complacently. She took a cigarette from her bag and Axel produced a lighter from his pants pocket. His chair was pushed back from the table; he sat with his legs wide apart and his arms stretched out before them. His fingers rested very lightly on his knees, and Clarisse half suspected he was involved in advanced isometrics.

"Were you at the party last night?" Axel asked.

Beneath the table Clarisse nudged the toe of her shoe hard against Valentine's thigh. She exhaled a waft of swirling smoke, and said, "No, I had to do inventory last night. Did you enjoy it?"

Axel nodded, and flexed his triceps diligently before taking another swallow of Drambuie. "Who was your friend then, Daniel—the Empress Wu?"

"A friend from Boston. And it was a man under all that paint."

Axel snorted approvingly. "Tell him he was pretty good. Fooled me. Is he . . ."

"Is he what?" prompted Clarisse when Axel didn't continue.

"Is he staying in town?"

"You want to get set up with a date?" asked Valentine.

"No. I was just wondering if he were still in town."

"No," said Valentine, not looking at Clarisse. "I put him on the commuter plane this morning. Still in drag."

Axel nodded distractedly, but seemed relieved. "Listen," he said, "you haven't seen Scott tonight, have you?"

Valentine shook his head. "Were you waiting for him out there?"

Axel nodded. "We had a fight—at the party, in fact—and he ran off. He took the car. I don't know where he went. We had reservations here tonight and I was hoping he'd show up. He didn't. I—"

Axel looked up at Clarisse and hesitated.

"You're not interrupting," said Clarisse. "Val and I have already exchanged confidences for the evening. In fact, I'm going to leave you two alone." When she stood, Axel started to rise, but she placed a hand on his shoulder. "No, please. Chivalry makes me break out in hives."

He stood anyway, and apologized. "I'm driving you away."

"No, no. I'm getting up at five in the morning, and still have to read the instructions that go with my alarm clock." She winked at Valentine and left the room, then stopped just outside the door on pretext of smoothing her skirt, and with satisfaction noted that Axel had already begun to spill to Valentine. "I have *no* idea where he's gone," she heard him say, "and I don't know what to do. I've—"

The rest was covered by Angel's yodel just outside the window.

As she stood outside the gate in darkened Kiley Court, Clarisse could hear laughter and splashing behind the high

65

hedges. Evidently there was someone in the pool. She unlatched the gate and went inside. All three wings of the house were dark, and a few candles in amber glass provided the only illumination. The waning moon was hidden behind clouds. She got all the way to the edge of the pool before she saw Ann and Margaret swimming in the nude.

The two women swam over. Their bodies were sleek and well-toned. They rested their folded arms on the tiles, and their legs gently paddled the water behind them. They looked like waterlogged cherubs.

"Hello again," said Ann with a smile. "Did you and Daniel have a nice dinner?"

"It was very pleasant," replied Clarisse with a smile. She was glad to see that most of Ann's earlier drunkenness had passed, or more likely been absorbed by dinner. And whatever disagreement between the women had caused Ann's tears in the Throne and Scepter had evidently been smoothed over as well. "And you two?"

"We ate at a place called the Forward Pass," said Margaret. "Did you know they have waiters dressed like cheerleaders?"

"I've heard."

"And," said Ann with emphasis, "they've got a great wine list."

Clarisse glanced toward the darkened wing of the house at her right. "Are Noah and Victor home?" she asked.

"Haven't seem 'em," said Margaret, wringing water from strands of her thick hennaed hair.

"Don't you want to swim?" asked Ann. "I love this pool, I love swimming in it at night. I love having it all to ourselves. I wish we could stay here the whole summer."

"No thank you," said Clarisse. "After the dessert I had tonight, I'd just sink. But I think I will sit out here for a few minutes, enjoy the night air, and try to think up a good excuse for not going to work tomorrow." She drew a chair up close to the edge of the pool, seated herself, and indolently lighted a cigarette with one of the candles. Ann and Margaret swam away and then back again when she motioned that she wanted to continue the conversation.

"I had a good time at the party," said Clarisse, looking down at the two women in the water.

"When my film comes back, I'll be sure to give you some prints of you and your friend." Ann paused and added, "You really are a beautiful woman." She reached for an opened bottle of wine that was beneath the table at Clarisse's side.

"Thank you," said Clarisse, smiling. She watched with interest as Ann poured out a full glass, spilling a little on the tiles.

"Are you living with Daniel or just staying with him?" Ann continued after Margaret pushed gently away to glide through the water.

"Valentine is . . . gay," Clarisse said, hesitantly. The candlelight provided not much illumination, but enough to show Clarisse what was in Ann's eyes. "But we, ah—"

"She's straight," shouted Margaret from the other end of the pool.

"*That's* what I meant to say," said Clarisse.

Ann sighed. "Who can be sure anymore? I mean, after Eleanor Roosevelt . . ." She shrugged. "Must be difficult for you in this town."

"Life is a trial," Clarisse admitted, then changed the subject: "Did you take a lot of pictures of costumes last night?"

Ann nodded. "Everybody was very nice about it."

"Did you get one of Cain?"

"Cain? Which one was he?" She swallowed off the glass of wine.

"The one with the *X* on his forehead, wearing a chiton."

"Is that like a tiara?" asked Margaret.

"No," said Ann quickly, "I didn't get his picture, and I'm glad too."

"Why do you say that?"

"Didn't you hear? Somebody killed him."

Clarisse made no reply.

Margaret swam up. "Yesterday morning," she said. "The tide was licking his heels."

"I know," Clarisse said at last. "I was the one who found him."

"Oh, that's terrible. Were you looking for him?" asked Margaret.

"It was terrible. Why should I be looking for him?"

"Well," said Margaret, "you know, we were all on the ferry together yesterday, and Ann and I saw you walking down the pier with him."

"That's right," said Clarisse. "He was trying to hustle a place to crash."

Ann threw one dark leg over the edge of the pool and hoisted herself out of the water, sitting sideways on the edge. Water splashed onto Clarisse's white shoes, and Margaret reached out of the pool to wipe them off with a towel. "Do you want a joint?" Ann asked. She took one from the pocket of her folded shirt and Clarisse leaned forward with the candle. Margaret, still in the pool, held on to Ann's feet under the water. She shook her head when Ann motioned toward her with the joint.

"It's not treated," Ann assured her, but Margaret still waved it away.

"Did *you* talk to him?" asked Clarisse.

Both women nodded. "But not on the boat," said Margaret. "Here."

"Here!" cried Clarisse.

"We got here about one-thirty yesterday afternoon. And after we had unpacked"—here Ann interrupted with a prolonged giggle that Clarisse had no difficulty interpreting— "we put on our suits and came out here. Shortly after that he came through the gate with his bag and said he was looking for his lover."

"His lover?" Clarisse repeated with astonishment.

"Well," said Ann, "I told him that we had just arrived, and didn't know who else was staying here, but that there wasn't anybody at home right then. So then he sat down and waited."

"But who could his *lover* be?" demanded Clarisse. "I know it's not Valentine. Maybe it was the man who was staying here last week."

"No," said Ann definitely to this last. "Terry O'Sullivan is my boss. I'd know if he had a lover—and he doesn't."

"He came to see Mr. Lovelace," said Margaret.

"Noah!"

"Jeff King sat right there where you're sitting now," said Margaret, "and just waited for about ten minutes. Then Mr. Lovelace came back. He was very surprised to see Jeff here, but he took him inside. They were in the house for about five minutes and then Jeff came back out."

"He was mad, too," added Ann with a gasp, after she had sustained a lengthy inhalation of smoke.

Clarisse sat puzzled and consternated for several moments. "I don't understand any of this."

"It's simple," said Margaret. "Jeff King had probably tricked with Mr. Lovelace a couple of times, and then blew that up into 'my lover.' Then he shows up on the doorstep expecting a place to stay, and Mr. Lovelace says he's all full up. So Jeff King goes away mad. That's bound to be what happened."

"Maybe," said Clarisse doubtfully. "Did you tell this to the police?"

Ann laughed. "Why? He wasn't killed here, after all." Then she added indignantly, "I'm not going to spend the last week of my vacation filling out police reports. Margaret and I have better things to do."

Now Margaret giggled.

Clarisse said nothing, but just sat staring across the dark courtyard toward her uncle's unlighted windows.

Chapter 12

Clarisse was reading Monday's edition of the local newspaper at breakfast when Valentine staggered in from his bedroom. He was wearing a pair of frayed gym shorts at least fifteen years old, and she was already dressed for work. While she poured him coffee and brought out doughnuts, he stared at the front page of the paper. The banner headline told of a rare species of whale found beached in Herring Cove; a minor drug bust was noted in the lower left-hand corner; a large center-page photograph showed a child sitting atop a pier fishing with a sunset shimmering behind her.

"The notice is on page six," said Clarisse. "Jeffrey Martin King, age twenty-eight, resident of Boston, leaving a mother and two sisters to grieve."

"And that's all?" asked Valentine, sipping his coffee.

"That's it. Mr. and Mrs. We're-on-Holiday-and-Don't-Want-to-Think-about-Anything-Nasty wouldn't want to hear about the thumb marks on his throat, and wouldn't want to know exactly where on the beach the body was found," said Clarisse sourly.

"I told you so," said Valentine, and pushed the newspaper aside, without bothering to turn to page six.

"I can't believe they'd ignore a *murder*. I've been through that paper three times, thinking I'd overlooked the article, but it's not there. I have a good mind—"

"Don't get started," warned Valentine. "There's nothing you can do."

They were silent for several moments. Clarisse sat across from Valentine and methodically tore the paper, page by page, into small scraps.

"So," she said at last, "you persuaded someone home last night."

Valentine nodded. "You heard us?"

"Not the part where you turned down the sheets, but everything else."

"It was Axel. We went to Back Street. I bought him a drink and he spilled his guts. Then I brought him back here and comforted him for about two and a half hours. Then we ran out of poppers."

Clarisse nodded toward the bedrooms. "He still asleep back there?"

"Left with the dawn. He started to get worried that Scott would come back and not find him home. Listen, why didn't you want Axel to know that was you in Oriental drag?"

"I didn't want to embarrass him. I saw him have that fight on the deck, remember? You didn't tell him, did you?"

"I didn't need to. He knew it was you."

"Why didn't he say anything?" asked Clarisse.

"He didn't want to embarrass you by catching you in an outright lie."

"Oh, well," she sighed. "Go back to the spilled guts."

Valentine took a sip of his coffee, thought for a moment, and then began to speak, in a sincere and deep-throated voice: "Dear Ann Landers:

"I am a thirty-seven-year-old swimming coach, moderately successful. I have seventeen trophies, a gold American Express card, a picture of me shaking hands with Lyndon Johnson, and a lover who is fifteen years younger than I am. I'll call him Scott DeVoto. Scott is jealous, insecure, and out of work. He doesn't understand me, he doesn't understand that just because I hop into bed with every man who winks at me that I still love him more than anything else in the world. I met him in the bliss of a coach-student fantasy. The fantasy faded, but Scott hung around. Which is fine, except that he's always afraid that I'll dump him for somebody who's got a better time on the hundred-meter freestyle.

"Three months ago I met a young man I'll call Jeff King. I saw him twice—it wasn't any big deal, except that Jeff King

71

gave me the clap, herpes, and crabs. Then Scott (my lover) accused Jeff (my trick) of stealing some clothes out of his closet. Ann, we nearly went to divorce court after that one. We had this huge fight in the VD clinic and it was terrible. It wasn't the first fight and it wasn't the last one, either. And the thing of it is, I still love him! Scott, that is, I still love him like I loved him on the day we first stared at each other in the showers. Am I afraid of growing old alone? Will I ever stop screwing around? Do I really enjoy public battles?" Valentine took a breath, adding at the last, "Signed, Anxious Axel."

"Love is a many-splintered thing," remarked Clarisse. "Do you think Jeff actually stole those clothes?"

"Axel says they were lost in the laundry," replied Valentine. "But who knows? I wouldn't put it past Scott to have thrown them out himself just so he could accuse Jeff."

"Axel supports that twerp, doesn't he?"

"Scott 'takes lessons,'" said Valentine with distaste. "Car repair, home cosmetology, computer programming, weaponless self-defense, sensuous massage, tai ch'i. He's not sure yet *what* he wants to do."

"When I think of the number of men in this town who are lover-subsidized . . . !"

Valentine nodded, and rose to pour more coffee. When he sat down again, Clarisse said, "Well, I had a little poolside chat with Ann and Margaret last night. They tried to induce me into a threesome."

"Were you polite in refusing?"

Clarisse sighed. "The approach and the denial were carried out with the utmost discretion and good taste."

"Tell me what they had to say."

Clarisse recounted the pair's assessment of Jeff King, concluding with Jeff King's astounding assertion that Noah Lovelace was his lover. "But that, of course, is not possible," she concluded.

"Why not?" said Valentine. "I'll bet you can't name Noah's last three affairs."

"Yes I can," said Clarisse. "Truck-Stop Betty, Butcher-than-Thou, and Amtrak Bob."

"Those aren't real names," argued Valentine. "How do you know that Jeff King didn't have a nickname as well?"

"You mean like the Cobalt Clone?"

"Or something. Anyway, it doesn't look like Jeff King was addicted to truth-telling—that man couldn't be trusted the length of his zipper. But if you're worried about it, why don't you just ask Noah?"

Clarisse replied uneasily, "The only time Noah and I have had together was at the party on Saturday night, and that was for only about five minutes."

Valentine eyed her. "Are you afraid you'll find something out?"

Clarisse didn't reply to this directly. "After the girls dropped their little bombshell, they went inside. I sat outside waiting for Noah to come home so I could ask him all about it. But the White Prince arrived first, and I said, 'Is Noah with you?' And the Prince said that Noah had gone to Boston early yesterday morning and wasn't expected back before tomorrow."

Valentine was puzzled. "Was it an unexpected trip?"

"I asked the Prince. 'He never tells me anything, how do I know?' he said." She glanced at the clock. "Oh, God, I'm late. The hordes will be beating at the door."

She rose and hurried upstairs to the bathroom. In a couple of minutes she came down again. "Listen, Val, I've had a thought," she said. "If you see the White Prince this morning, tell him to drop by and see me in the shop."

"You think he might have known Jeff King too?" Valentine said, and then looked thoughtful. "I wouldn't be surprised. Everybody in town except me seems to have screwed him—or been screwed by him."

"Don't tell him what it's about. Let me do a little fishing, Okay?"

The White Prince was tall, as thin as Jacqueline Onassis' little finger, and had immaculate, deeply tanned skin and classically regular features. His cheekbones were high and his eyes were green and cold. The White Prince's crowning glory was the shock of absolutely white hair that grew luxuriantly and with no sign of receding; the rest of his body was hairless. He couldn't even grow a mustache.

His history was vague and had something to do with Missouri and the wrong side of the tracks. Though his public manners were languid and refined to a degree more appropriate to a hedonistic Olympian god than an ex-schoolteacher living by his lover's grace, it was rumored that his mother paid for his piano lessons after a successful appearance on *Queen for a Day*.

He appeared in the doorway of the Provincetown Crafts Boutique an hour after Clarisse had opened the shop. His sandals were carved of birchwood and laced with ostrich-leather thongs. His drawstring pants were of raw cotton. His pink shirt of the same material was ornamented with tiny scraps of mirror sewn about the collarless yoke. Despite the intense heat of the day, a lightweight white sweater was thrown over his shoulders, with the arms loosely knotted just beneath his throat. His large sunglasses were rose-tinted and red-framed.

The White Prince always posed as if a candid fashion photographer were lurking behind every bush. Clarisse conjectured that he must have had a costumier, makeup man, and hairstylist just off-stage. The shop was filled with tourists, who stopped their browsing to stare. The White Prince took no notice of them; to him, heterosexuals were invisible.

"Daniel told me to stop in. Here I am." He advanced toward the counter, but stopped suddenly when the plastic whale in the barrel suddenly shot up a geyser of water. "Is that likely to happen again?" he asked.

Clarisse nodded. The White Prince gingerly lifted the whale from the water and dropped it onto the floor. There was a loud crack of plastic. "Oh, sorry," he said.

"Don't worry," said Clarisse with a broad smile. "It could have happened to anyone."

"Daniel said that you wanted to talk to me about Jeff King." The White Prince's voice was loud and quite conversational in tone; it was a voice that took no notice whatsoever of those standing all about them.

"Valentine gives everything away. He's got no subtlety."

"What did you want to know?" The White Prince had long ago given up wrinkling his brow, for fear the lines might then never go away.

"*Did* you know him?"

"Of course. Everybody knew Jeff King. He's been a regular in P'town the last few summers. Three or four years ago, he stayed here all summer long, and became the largest-scale dope dealer we had. Summers after that he'd come back every weekend or so, just to deal."

The customers in the shop gave up all pretense at browsing, and listened intently. The White Prince stood a few feet away from the counter—so that Clarisse would have a full view of him—and Clarisse saw that it would be pointless to ask him to use anything resembling a confidential tone and volume.

"What did he sell?" she asked.

"The usual—Black Beauties, meth, MDA, speed, 'ludes, coke. Things to get you up, things to keep you going, things to lay you out."

"Did you buy from him?"

"I used to. When he was still bringing good stuff."

"And he wasn't anymore?"

"No. A few years ago he couldn't have gotten away with it. P'town wasn't so popular then, *everybody* was a regular, and word would have gotten around fast that he was cutting his MDA with speed. MDA's just great—a good glow and you can have sex all night and do *everything,* if that's what you're into. MDA's the best thing that's happened to romance since they hung the moon in the sky. But once you start cutting it with speed, it makes things rocky—you'll still do everything, but you'll do it faster and harder. Lately he'd been coming back and selling to strangers, people who didn't know his drugs were no good. That may have been why he was murdered—"

There were gasps in several corners of the shop. Clarisse turned and smiled to everyone. "He wasn't a *close* friend," she assured them.

"Somebody got hold of his bad MDA, or his bad coke, or his bad meth, and tried to get their money back, and so on and so forth. You found him, Daniel said?"

"Yes," said Clarisse, and gave change for a ceramic clown nervously bought by an elderly woman in a puce blouse and matching synthetic pants.

"It must have been a terrible experience," said the White Prince languidly. "I heard that you threw him over your shoulder and carried him to the courthouse. His hands were scraping the asphalt. I heard that at five A.M., it looked just like *The Creature from the Black Lagoon,* and the meat rack scattered when they saw you coming."

"It wasn't quite like that," said Clarisse delicately. "Did Noah know Jeff King?"

"They knew each other, but they didn't get along. Trouble in the past, I think."

"What kind of trouble?"

"Ask Noah, I don't know. I never ask questions about the past," said the White Prince, tracing a recently manicured fingernail across his lower lip.

Chapter 13

When Valentine was still setting up the register that Monday afternoon, he noted in the mirror that Terry O'Sullivan had sidled onto a stool just behind him. The time for polite discouragement had passed. Terry had come to Provincetown with the intention of staying only a week—from Saturday noon to Saturday noon. But when, after a week of being hounded, Valentine at last agreed to go with him to the Garden of Evil party, Terry had taken a room at the Boatslip for another two days. And now Valentine was dismayed to find Terry was exceeding even *that* extension.

Valentine did not speak. He had once thought Terry O'Sullivan handsome: a compact well-defined body, dark features, short curly black hair and a full black mustache. But his polite pushiness during the week that they had lived in Noah's compound and the absurdity of Terry's assumption, on their first date no less, of some sort of relationship existing between them had effectively dissipated that attractiveness in Valentine's mind.

"Daniel, could I have my usual?" said Terry.

"What's your usual?" said Valentine.

"You know," Terry replied, puzzled, "club soda and lime."

When it was set before him, Terry made no motion to pay. "That's seventy-five cents," said Valentine shortly.

Embarrassed, Terry laid down the money. Valentine made change and then walked to the other end of the bar. He opened a *New Yorker* that was lying on the beer cooler and began to leaf through it.

"Daniel," Terry called, "come down here and talk to me."

Valentine slowly closed the magazine, and said pointedly, "I thought you were supposed to go back to Boston this morning?"

"I was," said Terry nervously. "I've got stacks of work on my desk, and they're starting to call me. They can't do without me," he added with feeble pride. "But I don't care. I've taken off two more days. I couldn't leave Provincetown until you and I had gotten the chance to really talk. I've been wanting to talk to you, Daniel, but I had to think it out first. But this is a perfect time now." He glanced around; the only other person in the Throne and Scepter was the waiter sitting in the open doorway. He pretended not to listen to the conversation.

Valentine waited.

"Really *talk,* I mean," Terry said in a low grave voice.

"All right," said Valentine grimly. "But I'm going to warn you. You may get answers you're not going to like."

There was a silence of some moments.

"I think we're good together," Terry began in a rush. "The first time I met you, Daniel, it just about blew my mind. I got this feeling like I've never had in my life." He paused significantly, but if he expected Valentine to echo the sentiment, he was to be disappointed. Valentine stood stock-still, hands folded across his chest, and looked closely at Terry. "At the same time I had the feeling—and I know it's right—that something good could come of it. If you'd only give it a chance."

Valentine said nothing.

"Oh," Terry went on after a moment, "I'm not asking for a commitment, it's too soon. I just want you to give *us* a chance. I've made reservations at the Boatslip for every weekend they had a room open. I won't be back next weekend but I will be the one after. And all I want you to do right now is say you'll set aside that weekend for *us.* That's all the commitment I want. Daniel, this could be the start of the most important part of our entire lives."

Daniel made no reply.

"You look angry," said Terry slowly.

"I am."

"How could you be angry?!"

"Because," said Valentine softly, "that's exactly what I *don't* want—a relationship, I mean." His gaze was harder than Terry O'Sullivan was prepared to deal with.

"Yes, you do want it," said Terry O'Sullivan, glancing away. "But you're afraid of making a commitment."

"Listen to me." Valentine's voice was icy. "If you will remember correctly, you and I have never had sex. We occupied the same bed for an hour and a half, while you talked. I didn't even get to take off my cufflinks. And you know what else? That was it. That was the high point of what you consider 'our relationship.' Because that's as far as you're ever going to get with me."

Terry was crushed.

"I would have had sex with you—but I get you into bed, and you pull out this contract you want me to sign. I have no interest in contracts."

"Gay people ought to learn—"

"I'm not talking about 'gay people,' I'm talking about *me!*"

"What about Clarisse?" Terry demanded pettishly.

"Clarisse," said Valentine solemnly, "is the love of my life."

"She's a woman!"

"You're being rude, Terry."

"Rude! I *love* you. I . . ." He broke off in frustration.

"Please leave," said Valentine quietly, not allowing the man to speak.

"No, you're wrong, I—" Terry began again apologetically.

"Please leave," Valentine repeated in a tone of voice that wasn't as soft as before, "because if you don't leave now all you'll have to show for your efforts is an early grave and all I'll have is a cell in Walpole."

Terry eased off the barstool. "You do this a lot, don't you?" he said bitterly. "You must, 'cause you're real good at it."

Valentine turned toward the cash register and pushed several buttons in rapid succession. In tiny red lights across the screen were spelled out the words REGISTER CLOSED.

Terry O'Sullivan turned on his heel and left the bar.

Chapter 14

On Monday night Clarisse decided that she ought to catch up on a little sleep. She had gone to bed not at all on Saturday night, slept little on Sunday, and after two full days of work and thinking about a dead man, she was weary. She declined an invitation to dinner with Valentine, and dined alone on a glass of red wine and half an Explorateur cheese—her favorite. After a leisurely bath, she put on a fresh nightgown, slipped between the sheets and was asleep within five minutes.

Next morning she awoke ready to face life. She checked on Valentine and discovered that he hadn't returned home the previous night. She made coffee for herself, indulged in part of an Entenmann's pecan Danish ring, and sat down at her makeup table still with plenty of time to get to work. Provincetown, she reflected, *could* be very pleasant. Birds sang in the coffee tree.

She heard the gate rasp open. Thinking it would be Valentine returning after a night's successful hunting, she went to the window to greet him. It was Noah carrying a suitcase. She called down and waved.

"You're back!"

He looked up at her and smiled. "Yes. Come on over and watch me unpack."

She hurriedly dressed and ran across the courtyard. She opened the door of Noah's apartment, and the White Prince lunged at her down the hallway with an Electrolux. He wore

80

his birch-heeled sandals, which clattered noisily—audible even above the vacuum cleaner—on the bare floor, white silk designer shorts, a Kelly green tank top, and a dozen thin gold bangle bracelets on each wrist. A white rubber skullcap with green stars protected his hair against flying dust.

"Hi!" he screamed over the vacuum cleaner. "Noah's upstairs!"

Clarisse jumped out of his way, slipped past, and went up the stairs. Behind her the vacuum cleaner was shut off, and the Prince shouted behind her, "Did you steal my Bon Ami?" He gave it a French pronunciation. "I can't find it anywhere."

"No!" cried Clarisse, and knocked on the door of Noah's bedroom. The vacuum cleaner started up again down below.

Noah opened the door, motioned her in, and closed the door behind her. The sound of the White Prince in the throes of housecleaning were mercifully dampened.

Clarisse looked around the room. It had been redecorated since she'd seen it last. The walls were deep rose, the prints were Japanese, the furniture was lacquered black—but the windows still looked out on unmistakably sea-resort foliage and Cape Cod sky. Noah's bag lay opened on the quilted black bedspread.

"I thought you hired someone to do the cleaning," remarked Clarisse. Outside the room, they could hear the Electrolux being dragged up the stairs.

"I do. But the Prince took the wrong pill this morning. What he thought was vitamin E was actually an Eskatrol. And at ten o'clock on a Tuesday morning what is there to do on an upper except clean house?"

The White Prince flung open the door of Noah's room and demanded, "Where is the drapery attachment for this machine? Who took it?"

"All the attachments are kept in the bathroom closet," said Noah patiently.

The Prince rattled his bracelets and slammed the door.

Noah shrugged, and began sifting through his bag for dirty clothes. He looked up and smiled at Clarisse.

"Where did you go?" she asked.

"Boston."

"I didn't know you were going."

"I didn't think to tell you. I had planned the trip. I went to

Boston to look over the Brookline Swiss Miss, and to have breakfast with Cal. You know Calvin Lark, don't you?"

"Yes," Clarisse replied. "His firm also represents the real estate office where I used to work. And naturally he's a friend of Valentine's. He's the one who suggested that I go to the Portia School of Law."

"Well," said Noah, "I went to see Cal. On business."

He closed the drawer on some shirts that hadn't been worn. "Why so curious? Did I miss something around here?"

"Yes, as a matter of fact. Somebody at the party on Saturday night was murdered."

"I know."

"And I found the body."

"I know."

Clarisse turned her head a little and raised her voice; the vacuum cleaner had started up again, this time right outside the bedroom door. "How did you know? You left town so early after the party."

"The Prince told me, I—"

The Prince opened the door. "I need another extension cord," he cried. "I want to reach out the window and clean out the eaves."

"Clarisse," said Noah, "I'll be right back."

He left the room with the White Prince directly behind him, volubly complaining that not only was their only dust mop in terrible shape but that *somebody* had bought AmWay furniture polish instead of good old Lemon Pledge.

Noah was gone for nearly ten minutes. Clarisse peeked into his bag. When he returned he apologized for having taken so long, and said that the Prince had given him a long list of supplies that *had* to be purchased before he could get *anywhere* with the cleaning. "So I'll have to go out now. Do you want to wait here for me?"

"No," said Clarisse, "I have to get on to work." She couldn't help but feel that her uncle was doing what he could to abandon the conversation regarding his trip to Boston. "Do you want to have lunch?"

"I wish I could," he replied. "But I'll be at the restaurant. On Tuesday, we have the management-union meeting. Angel and I stand up against the wall, while the twenty-five waiters hurl complaints and day-old pastries at us."

"Maybe tomorrow then?"

"We'll see," he smiled. "But don't worry about it, we're both here all summer."

She stood to go. "Well, why don't you walk me to work, I—"

"Opposite direction," he smiled with a small sigh.

They at least went down the stairs together, and out into the courtyard.

The White Prince leaned out the bathroom window and screeched, "Don't forget the brass cleaner!"

Chapter 15

Clarisse arrived at the Throne and Scepter just as Valentine was finishing his shift that evening. He was preoccupied and morose.

"You broke a heart today, didn't you?" said Clarisse. "You're always like this when you have to break a heart. Whose was it?"

"It wasn't today, it was yesterday. And I'm still depressed." Valentine told her about the disagreeable scene with Terry O'Sullivan.

"I would have thought you'd be over it by now. Didn't you go out last night?"

"Yes."

"And didn't you find someone to make you forget your woes?"

"Yes. But today my woes has come in five times to beg forgiveness."

"What you need," said Clarisse sympathetically, "is a little sign over your bed: *Three-Day Limit*. For your birthday I'll make you one, in needlepoint."

"What I really need," said Valentine, "is about five more drinks."

"Sorrows float in liquor," said Clarisse. "Only two things help in a case like this. Spending a lot of money or trying on a lot of clothes. I've been known to do both."

"I don't want to do either."

"Accompany me to Maggie Duck's Duds," said Clarisse. "I've got to find something for my date with Matteo."

"Who?"

"Matteo Montalvo—the object of my most recent flights of fancy. Cops, as you must know, have a fine eye for the details of a woman's clothing, so I have to be careful what I wear."

"Maybe that's the answer," said Valentine.

"What?"

"Maybe I should date straight men—at least they wouldn't be prone to fall in love. And I *dote* on uniforms."

"Val, I have just secured for myself the only good-looking unattached straight man in this town. If you try to take him away from me, I will staple your ears to your shoulders."

Maggie Duck's Duds was two doors down from the Swiss Miss in Exile. Its stock was good used clothing from the middle forties through the late fifties, and considering the general run of Provincetown markups, its prices were moderate. At nine o'clock on Monday night the shop was not crowded, and Valentine and Clarisse were alone in the back room. From a rack of dresses Clarisse selected a ocher silk evening gown with silver bugle beads shot through the pleated bodice. Valentine was desultorily examining shirts.

Clarisse put on the dress, then stood before the mirror and adjusted the wide padded shoulders and bolero sleeves. She turned around to admire the sway of the floor-length skirt, then pulled back her hair, pushed it up, lowered her eyelids and pouted, affecting her sultriest demeanor.

"For once you got it right," said Valentine over her shoulder in the mirror. "Miss Barbara in *Double Indemnity.*"

"Wrong," she said, "Diana Dors in *Yield to the Night.*" She shook out her hair and turned to face him. She stopped abruptly, staring.

"It's that bad?" he asked. From the rack against the wall, he'd put on a shirt with a scattering of black figures on a vibrant red background. The price tag dangled from one of the sleeve buttons. When Clarisse continued to stare dumbly at him, he stepped in front of her and examined the shirt in the mirror. "Collar too wide?" he suggested. "Cut too full?" He pulled the sides of the shirt close to his body. "Maybe cleavage would help," he said, undoing two more buttons. "Clarisse, say something!"

Clarisse faltered. "It . . . looks good on you too."

"What do you mean, me too?"

"I mean that it looked great on Jeff King when he got off the ferry wearing it."

"Oh," said Valentine softly, after a moment, "dead man's clothes . . ." Then: "Are you sure it's the same one?"

"Tulips," she said, poking at one in the design. "Black tulips on a field of red. When I talked to him after we got off the boat, he was wearing *this* shirt. Bend down."

Valentine leaned forward, and Clarisse pulled back the collar. "No name tag, but the label's period. I'm sure it's his. I'd bet Richard Nixon's political future that there's not another one like it on the eastern seaboard."

"Maybe you're right. Think anything else of his turned up here?"

"Let's look! What should we look for?"

"Underwear," suggested Valentine. "Drug dealers always wear fancy underwear. Oh, and Jeff King went swimming Saturday afternoon, look for a wet bathing suit."

"You're making fun of me."

"Of course I am. The only things you saw him wearing was this shirt and the toga on Saturday night."

"Chiton," Clarisse corrected.

"All right, *chiton*," Valentine allowed. "So he probably got buried in the chiton, and now you want to look through an entire clothing store for the rest of his wardrobe, which you never even saw?"

"This is a clue!" she said, stabbing at another tulip.

"Wouldn't it make more sense to ask the people running the shop where *they* got the shirt?"

"I suppose," said Clarisse, reddening a little. "Take it off." While he did so, she went into the dressing room, and removed the gown. She came out with it over her arm. "Get out your Visa, I don't have enough money for this."

At the counter Valentine surrendered his plastic. Clarisse smiled at the young woman behind the counter. She had a pinched, heavily rouged face, pouting lips outlined in fuchsia lipstick, chopped hair dyed black and streaked above the ear in turquoise. She wore a leather jacket so tight that she probably could not have raised a cigarette to her lips, and whether she would be able to handle the cash register was itself an interesting question.

"You have wonderful things here," Clarisse remarked. "Where do most of them come from?"

"Here and there," the woman replied without any interest at all.

"Do you order from Boston and New York, or do you get most of it from around here?"

"Both," replied the young woman, snipping the price tags from the shirt and the gown.

"I mean," said Clarisse carefully, "do you have buyers who go out looking or do you buy lots of clothing or does some of it get donated?"

"You looking for a job?" asked the woman behind the register warily. "We don't need anybody."

Valentine sighed and held up the shirt. "This shirt is evidence in a murder case. It belonged to the drug dealer who was strangled on Saturday night. We're buying it to take to the police. Can you tell us where you got it?"

The girl brightened, and smiled at Valentine. "Oh, sure, I remember that shirt. A whole bag of stuff got left out on the front steps last night, but that shirt was the only thing decent in it. I went through the bag myself. People leave stuff here sometimes. They sort of think we're a charity 'cause we handle used clothes."

"Do you have the other things here?" asked Valentine.

The girl shook her head. "Everything else in the bag was new. So was the bag. It all just got dumped."

"What else was in the bag?" demanded Clarisse.

"Oh, just stuff," replied the girl vaguely.

"What sort of stuff?" asked Valentine with a smile.

"Well, there was some underwear and socks, and some black tennis shoes. I would have kept those, except they were too big for me. Combs and brushes and shampoo and all like that. Pair of jeans, button-fly Levis. Oh, yeah, I kept those too. They're in back, but I don't remember what size they are. Everything gets washed first—disinfected too—before we put it on the racks," she added reassuringly.

Clarisse's gown was wrapped, but Valentine insisted on wearing the shirt. He winked at the girl behind the counter, and pulled Clarisse out of the store.

"You're the worst interrogator I ever saw," complained Clarisse.

"No I'm not," protested Valentine. "You weren't getting anywhere with that girl."

"She was after your body. She would have given away state secrets for a chance to hold your hand. I think you ought to go back and try to find that bag they threw out. There could be a vital clue in that bag."

"I am *not* going to spend a warm summer evening sifting through a dumpster. Besides, everything got picked up today."

"All right, all right. I wish you wouldn't wear the evidence though—I think it's bad luck."

"I *like* this shirt," said Valentine. "And it is *not* evidence. It's just a shirt."

"Who do you suppose left that bag on the doorstep of Maggie Duck's?"

"The killer," shrugged Valentine. "Who else?"

Chapter 16

"Where shall I take you to eat?" asked Valentine, pausing to admire his appearance in the dead man's shirt, as he saw his reflection in the mirror of a shop window.

"I had wanted to eat outside, but it's too damp now. The fog's coming in, so we might try—" She squeezed Daniel's arm, and whispered, "Well, look who's back together again." Coming toward them down Commercial Street were Axel Braun and Scott DeVoto. The two men made an undeniably handsome couple, talking and laughing quietly together, dressed in worn levis and black tank tops—outfits that spelled *we're going dancing*.

"All is forgiven," said Valentine. "Thank God. Don't stop to talk," he warned Clarisse. "I don't think Scott's even above jealousy of a woman."

The men did not see Valentine and Clarisse until the couples were no more than twenty feet apart; they met just in front of the well-lighted windows of an antique shop.

Axel's face registered pleasant surprise, but Scott's smile withered. As Valentine and Clarisse drew nearer, the two men's mouths dropped open together. They stopped dead.

"Hello," said Valentine, and despite his own injunction paused on the sidewalk.

Axel nodded a distracted greeting. Scott drew in his breath sharply between clenched teeth: it was an inadvertent hiss.

Valentine and Clarisse glanced at each other apprehensively.

"Where did you get that shirt?" demanded Scott.

"Oh, do you like it?" cried Clarisse. "I just picked it out for him not five minutes ago. We—"

"I don't believe you!" cried Scott in livid anger. He turned to his lover. "It's one trick too many, just one fucking trick too many! And you told me you had never even seen him before!" He stalked off, leaving Clarisse and Valentine paralyzed with wonder.

Axel turned to them in agitation. "I bought Scott that shirt the first Christmas we spent together. There's not another one like it." He grimaced. "It's one of the shirts that Jeff King stole out of Scott's closet."

With this brief explanation, he hurried off to catch up with Scott. In another moment the voices of the two men were raised loudly and with such vituperation and heat that a number of people paused to listen. Valentine and Clarisse watched as Scott jumped up onto the raised veranda in front of the Throne and Scepter. He was alone on it as on a stage and, as any good actor would, he stood in the spotlight. When he waved his arms, the encroaching evening fog swirled in the bright yellow light.

"Whore!" Scott screamed. "I could kill you! Tricking in the morning, tricking in the afternoon, tricking every time I go to take a leak! What disease will I get this week? And now you give away my clothes to your fucking tricks! You want this shirt too?"

Axel Braun had climbed the brick steps at the end of the veranda, and was slowly approaching his lover. Axel said something, but his voice was so soft that Valentine and Clarisse could not hear the words. With an agonized cry, Scott DeVoto ripped off his shirt.

"Nice chest," remarked Valentine, "I didn't realize it was so hairy."

"Take it!" shouted Scott. "Find some fucking trick, and give it to him!"

"That shirt would look great on me behind the bar," said Valentine.

"Val!" cried Clarisse, "have you no compassion for the misery of the betrayed?"

"Not when misery is staged to such effect," replied Valentine nodding toward Scott, who was surreptitiously glancing over the street to see how much attention he had garnered.

"Come down," said Axel, and took Scott's arm.

Scott jerked away, losing his balance and tumbling off the veranda onto the sidewalk. His fall was broken by a passing mob of teenaged girls with ice cream cones; they shrieked and scurried on. Scott sat, weeping convulsively, with his back against a lamppost. Passersby on the street regarded him with more curiosity than pity. Axel stepped down from the veranda, lifted Scott up, and spoke earnestly to him for several moments. Across the street, Valentine and Clarisse lit cigarettes.

Scott nodded dumbly to whatever it was that Axel was saying, and in a little while he took back his torn shirt and they moved off together toward the A-House. They were quickly swallowed by the mist, which had now grown quite thick.

Valentine and Clarisse resumed their interrupted progress. "Does Axel *like* these scenes?" she asked.

"He says he doesn't," said Valentine, shrugging vaguely. "But it's the third public altercation they've had since the party."

"You've had quite a day," said Clarisse. "Yesterday Terry O'Sullivan provided the dramatics, and tonight it was Scott DeVoto. What power do you have that induces perfectly reasonable homosexuals to give performances that would reduce Tallulah Bankhead to a rank Method actress?"

Valentine shrugged again, and they proceeded in silence. Traffic crept along the crowded street at its usual pace. Tourists thronged the shops, and gay men had begun to emerge from their guesthouses for an hour's stroll and dinner before they went out to the bars. Even underneath the sweet dampness of the fog they could smell frying fish and pizza and cotton candy. On one side of Commercial Street a great crowd blocked the sidewalk in front of the booth of a street artist doing the portrait of identical twins who looked to be about seven or eight years old, and on the other side a long line was forming before the movie house for the opening night of the Diana Dors retrospective. Valentine and Clarisse overheard conversations and arguments, they saw assignations being made, and watched drugs and money changing hands. There were spurts of laughter and petulance. Friends of Valentine grinned and waved to him from shop and restaurant windows. As they passed the Provincetown Crafts Boutique, two women emerged clutching identical wooden fishermen and their vibrant red lobsters. There were noisy crowds on the sidewalks, and indolent crowds at the outdoor restaurants, and distracted crowds in the shops and the cafes—and no one thought of Jeff King, whose corpse lay on ice at the undertaker's, three doors from the bar where he once scored three thousand dollars' in cocaine in a single evening. The fog thickened by the minute, roiling in from the bay, cloaking everything. Every few feet, faces grinning or grim appeared suddenly before them, and then quickly disappeared behind.

"God," said Valentine evenly, "this place really is a garden of evil."

Chapter 17

They proceeded through the fog to the Cafe Blasé, but because of the dampness and because the waiters would have been able to find them outside only with the assistance of a flashlight, they ate indoors. From their table beside a second-floor window they could see neither the adjoining house, about seven feet away, nor the ground, about ten feet below. They talked of the relationship between the swimming coach and his untrusting lover, and of the general turbulence that seemed to affect all such pairs spending time in this particular resort.

"Despite the decals that I sell by the dozen," said Clarisse, "Provincetown is *not* 'For Lovers.' "

"Maybe you should make one up yourself that says: 'Provincetown: Trick Heaven.' "

After dinner, like the blind, they stumbled through the fog to the Back Street Bar, where Clarisse paid Valentine's dollar cover and pushed him gently down the stairs. "What you need," she said, "is a stranger who's never even heard of Terry O'Sullivan, and somebody who doesn't care that you broke somebody's heart. Find a man who's happily married, and only wants you for your body."

Clarisse was right, of course, and in Back Street, Valentine was determined to throw off the pall of guilt that he knew Terry O'Sullivan had dexterously and deliberately thrown over him. He ordered a beer and wasn't displeased to find that he was handed two bottles, it being two-for-one night.

He wandered around the bar, not quite as crowded as it would be in another half hour, and was surprised to find, in the darkest corner of the innermost room, Axel Braun. He was without Scott, but he was not alone.

He was in close smiling conversation with a man Valentine's age, Valentine's build, and Valentine's coloring. Valentine, in fact, knew the man.

He went over, nodded to both, said, "Hello, Jimmy. Hello, Axel. Axel, could I speak to you for a moment?" Without waiting for a reply from either of the men, he drew Axel aside. Axel didn't even have time to pick up his beers.

"I was just about to pop the question," said Axel, miffed.

"I know," said Valentine. "That's why I got you away from him."

Axel stared at Valentine with puzzled eyes and with a little hostility. "Why?"

"That's Jimmy the Ripper," said Valentine.

"His name's Jimmy—"

"He'll take you home, and he'll rip all your clothes off—you won't have any buttons left. He's the only man in Provincetown who can tear a hole in new denim with his fingernails."

"Could be hot," mused Axel, and glanced back at the corner where Jimmy sat scowling.

Valentine gave Axel his second beer. "He rips all your clothes apart, and then he throws you out."

"Oh."

"Now, if you want to go back, you can. I don't want to rescue a man who prefers to drown."

"I guess he's not a cuddle-bunny then," said Axel with a sigh.

Valentine shook his head.

Axel grimaced. "I didn't want to be alone tonight."

"The making-up with Scott didn't last very long."

"No," replied Axel, "we didn't even make it to the door of the A-House. He's there, I'm here. And an hour from now he'll come over here—or I'll go over there—and we'll glare at each other across the dance floor. Then one of us will go over, and then the other one will apologize, and then we'll go home together—and we'll have another fight before we get to the front door. Then Scott will cruise the beach for the rest of the night and I'll have the double bed all to myself. So I was in here trying to pick somebody up so that when Scott came over here he wouldn't find me."

"I'll hide you," said Valentine.

"Would you?" said Axel, winking solemnly.

After seeing Valentine through the basement door of Back Street, Clarisse had sat down on the brick balustrade and smoked a cigarette until a taxi stopped before the bar to discharge someone. She commandeered it and returned to Kiley Court. The taxi inched through the fog, and Clarisse sighed in the back seat with a delicious melancholy that she intended to enjoy, deepen, and prolong.

Ann and Margaret were once again in the pool, but she would not have known that except for the sound of their splashing.

"Only the hopelessly in love," remarked Clarisse to the women cloaked by the fog, "would be out swimming on a night like this."

"It's *so* romantic," said Margaret, bobbing up through the mist into sudden visibility.

"And I'm *so* stoned," said Ann, invisibly.

Clarisse went inside, climbed to her bedroom, and sat at her vanity. By candlelight she set just the front of her hair in the smallest perm rods she could find. She then climbed into bed and read a few chapters of a paperback reprint of *Laura*, then fell gently asleep, lulled by the distant splashing and soft voices of Ann and Margaret in the pool below.

Her sleep was disrupted once during the night, when she dreamed of money and brown sugar. She rose from bed without turning on the light, stepped on *Laura*, and groped her way down the hall to the bathroom. She drew a glass of water, and as she drank looked out the window into the courtyard. The fog was there, but not as thick as it had been earlier. She still could not see the ground, and Noah's wing of the house looked like the Giant's castle, built in the clouds; the ivy and roses were the beanstalk that Jack the Giant-Killer would climb. Clarisse was pleased with this simile, which she thought rather good for the middle of the night.

But the glass trembled in her hand, and water spilled over her nightgown when she saw that very figure of Jack the

Giant-Killer appear suddenly in the middle of the courtyard. A naked man—at least his chest was bare, in the misty shadows she could see nothing below his waist—looked quickly around and up at each of the three parts of the house, then strode toward the alley gate.

Without hesitation, Clarisse ran back to her room—tripping over only one object, an ottoman—to the window that looked out over the alley. She heard nothing, saw nothing.

She had stood watch for several moments trying to puzzle out the man's identity, when a movement in the bushes across the way caught her attention. The man had evidently hidden himself in the hedge around the garden of Poor Richard's Buttery. But Clarisse gasped, for in no more than ten seconds, he apparently had dressed himself—and as he departed, his boots sounded angrily on the gravel.

Clarisse returned to bed, and awakening in the morning she recalled the incident only vaguely. She reflected how many men were on the streets late at night in Provincetown, and how easily one or another of them might have wandered down Kiley Court.

Having had little to drink the night before, and having gone to bed earlier than was usual, she awoke before her alarm, and made a happy resolution that she would never take another drink or see the other side of midnight again in her life. This she forthwith emended to exclude the coming evening, that of her date with Matteo Montalvo. She wondered if policemen were like firemen—she had handled a couple of those.

She showered, pulled the perm rods from her hair, then meticulously fixed the loose curls into a Betty Grable poodle. Discreet lipstick and eye makeup were applied and then she donned a coral cotton dress with white gardenias worked into the elasticized bodice. She picked at the puffed sleeves of the dress as she gazed out the window and marveled at the splendor of the morning.

She adjusted the straps on her heavy-heeled sandals, and was about to descend the stairs when she decided instead to rouse Valentine to have coffee with her. Her ulterior motive in this was to show him that at least *once* she could rise before him. His door was cracked and she gently pushed it open.

She was consternated to find that he was not alone.
Valentine lay with his back to her, his face buried in the
pillow, one arm thrown across someone's broad chest. Clarisse stretched up on her toes to identify the other man, and
sighed loudly when she saw it was Axel Braun. He turned in
his sleep and drew his arms tightly around Valentine's back.
Valentine moaned contentedly, and nestled his head in the
hollow of Axel's shoulder; he had evidently found the man to
make him forget Terry O'Sullivan. She meditated a few
moments on the mutability of human emotions. The reconcil-
iation of Axel Braun and Scott DeVoto she had witnessed the
night before had evidently not taken. She sighed again, and
Axel stirred at that noise. Clarisse withdrew quickly.

In the kitchen she fixed fresh coffee, squeezed oranges for
juice, and spread a generous amount of butter across a heated
scone. She set the dishes on a round wicker tray, and with her
prescription sunglasses plunked just behind the poodle, went
out into the courtyard.

The sun was blindingly bright against the upper walls of the
compound, but the courtyard itself was shaded by the coffee
tree. She pulled up a chair and seated herself, after setting the
tray on the table beside her. In putting down the tray she
inadvertently smashed one of the glass candleholders on the
flagstones, but no accident so minor could destroy her mood
this morning. As she shook out her napkin in her lap two gulls
shrieked overhead; she put on her glasses so that she might
see them better. The salt air was refreshing and happy. She
watched the gulls circle the house, and wondered what made
them swoop so low.

She smiled, rubbed her chin, and glanced up at Valentine's
window wondering why she had not heard his usual thrashing
about the night before, looked at the flowers in their beds,
sniffed their damp fragrance, and was just about to take the
first bite of her scone when she chanced to look into the pool.

Bobbing face-up in the water beneath the diving board was
Ann, her hair tangled about her forehead, her arm caught in
the drain trough and her sightless eyes staring blankly up into
the cerulean sky.

PART III

Rhythms and Blues

Chapter 18

The official Provincetown season runs from Memorial Day to Labor Day. Though the three months before gradually build up to that terrific crush, and the three months after are needed to wind down from it, that short season provides more than three-quarters of the municipality's annual income. Perhaps only January and February find Provincetown really quiet. As might be expected, the full-time residents say that is the sweetest time of all.

During the summer, the town, whose permanent population is only a little over four thousand, will play host to nearly twenty-five thousand each weekend. A holiday weekend will bring twice that many. A substantial portion of this crowd is homosexual. A peculiar and strict rhythm marks every day of the season. The weekend has the same schedule as the weekday, only the crowds are more intense and the hilarity louder. The Provincetown day begins before dawn when the Portuguese fishermen, most of whom live in the West End, rise and take to their boats. The keepers of tiny necessity shops, unpretentious restaurants, and small groceries open their shops between seven-thirty and nine. Day-tourists begin arriving soon after. Vacationers who have taken rooms in Provincetown's innumerable guesthouses, inns, and motels rise about ten and shuffle off to the beach, usually one of the wide, white, spectacular Atlantic beaches—five miles or so distant—rather than the narrow crowded fringe of gray sand that borders Provincetown Bay. The specialty shops and second-class restaurants open about eleven, and these are principally staffed by gay people who live in Provincetown throughout the season. Also about this time the gay people who have come for only a weekend—or a week or two or three—make a dour-faced, blinking appearance on guest-house verandas and decks, tell one another how much they drank the night before, and describe to their friends how hot the number was they got, or almost got, or had stolen from them at last call. After a light meal—overpriced and served with practiced condescension—they go over to the Boatslip, where they sit in the shade or lie in the sun or move from one to the other with the display of indecision usually reserved for Saturday night's outfitting. Still not fully awake, they stare vacantly about for a few hours, talk with their friends about what it was like last night as opposed to the night before, this week as opposed to last week, this season as opposed to last, and look over the crop of new arrivals. There is little cruising. In the town, during daylight hours, one suns and drinks and recovers from the night before—there is no sex, no hint of sex, not even the possibility of sex. If you want sex during the day, then you have to look for it in the dunes. In Province-town proper, sex, like heavy drugs and the stars, is reserved for the night. At five or six there is dancing—but only at the Boatslip, and invariably somebody is thrown into the pool with all his clothing on. At seven everyone disappears again to change for dinner; by the time they emerge from their guesthouses the ranks of the straight tourists have begun to thin, and the first-class restaurants have opened. In the twilight you sit with your ex-lover and his new lover and your new friend whom you met last night and his old friend he did not know was in Provincetown and the two women who have the room next to yours in the guesthouse, and a boy they've known for years who is very dear to them. You tell them where you live in the real world and what you do for a living and what famous people you have touched. You boast how long you've been coming to Provincetown and compare it to Key West and Fire Island. Then you carry your drinks to a table and get served by the man you tried to pick up the night before at Back Street, and you hope he's sorry that he didn't say yes now that he sees how many friends you've got and hears how witty they are. After dinner you go back to your room, smoke a joint, and wonder whether you shouldn't save your last Black Beauty for Saturday night. You walk around in your new cowboy boots to see if they pinch, and constantly peek out the door to see if the bathroom is free. Along about eleven you go out to the bars. For two hours the dancing and cruising is frantic, then everything shuts down and several thousand men and women whose adrenaline levels are just about at their eyebrows are swept into the streets of a New England fishing village with nothing to do and nowhere to go. Several hundred eventually end up at Dodie's pizza parlor, and hundreds more at a similar place on the other side of the courthouse. But there is still the parade up and down Commercial Street, almost entirely gay now, chatting and laughing and dealing drugs and making offhanded sexual propositions. The benches in front of the courthouse are a spectrum of leather and fey, loose sweaters and careful denim. In the next several hours, men wander the beaches or walk the streets hoping to be invited to one of the parties they can see through the lighted windows of guesthouses and private homes. And finally this downbeat close of the night coincides with the morning of the Portuguese fishermen, rising to their nets.

It is a scheme as formal and as rigid as the nine-to-five world so many of these men and women have come to Provincetown to escape.

It was a scheme which Valentine took in his stride, but which Clarisse was at pains to avoid. She never accompanied Valentine to the bars when he was out merely to cruise. She would not go alone to the straight bars in town because they all had the air, the smell, and the clientele of the confirmed drunkards' saloon. So, Matteo Montalvo's request for a date seemed like divine intervention in what might otherwise have been a summer devoted entirely to Valentine's company and the White Prince's well-stocked library of forties fiction.

What Clarisse had not counted on was spending the summer stumbling across the lifeless forms of recent acquaintances.

Chapter 19

Clarisse gaped at Ann's corpse, thinking *I never even learned her last name*. The two gulls that had at first drawn her attention to the pool swooped down again, and now Clarisse, appalled, knew why. She flung her untasted scone at the birds and rose from the chair and flapped her arms. The gulls wheeled away.

She realized that she must do something. She looked at the doors of the three apartments of Noah's compound, as if some prize awaited her behind one of them. Which should she try first?

She could not understand why Ann should be alone. Ann and Margaret had been inseparable; how could one have drowned without the other's knowledge? In any case, Clarisse determined to enter the middle apartment first. Her frantic voice calling to any of the men—Valentine and Axel, Noah and the White Prince—would draw Margaret to one of the windows and from there, looking down into the pool, she would see her friend's corpse. It was a sight Clarisse wanted to spare her, so she stepped quickly around the pool, not looking again at Ann's staring face, and slipped silently inside.

She called Margaret's name softly, then more loudly. She knocked on the frame of the door, then smacked the palm of her hand against the banister rail. She mounted the stairs, and continued to call Margaret's name. Suddenly she was afraid.

But Margaret was not in the house, and when Clarisse

realized this, she could not be certain exactly what her fear had been.

She moved to the wall that she knew was common with Valentine's room on the other side, and kicked it several times. Then she ran downstairs, out the door, and into her and Valentine's place. Above, she heard Valentine's inarticulate growl. She called, asking him to come down. When he protested, she cried, "Daniel! Look out the window!" When she addressed him as "Daniel," he always knew something of importance was afoot.

Without waiting for his reaction, she ran outside again and toward her uncle's apartment. Her commotion had already roused Noah and the White Prince, and they appeared simultaneously in the windows of their bedrooms, Noah bare-chested and the Prince in an emerald-green silk robe agitatedly running his fingers through his hair.

"Clarisse," called her uncle, "what on—"

She extended her arm toward the pool, as a ringmaster might introduce a new act. Hearing exclamations of surprise and dismay on all sides, she dropped into a chair and burst into tears.

Valentine had called the police and they arrived only a couple of minutes after Axel had departed—everyone agreeing that his presence might only further confuse an already complicated household situation. Along with the police came the curiosity-seekers: the staff of Poor Richard's Buttery, inhabitants of neighboring houses, and those who thrill to the sound of a police siren. The White Prince and Clarisse stood guard at the gate, and there probably wasn't one person in all of Provincetown who could withstand their withering looks and sarcasm.

The police were exasperated to learn that Margaret, who had evidently disappeared sometime during the night, had left behind neither her surname nor her address. Noah patiently explained that she had been only a guest of the dead woman, and that the house had been leased for the week to Ann Richardson. Margaret was from Toronto and lived with a woman named Joyce, but that was all anyone knew.

Clarisse had returned home at half past ten, and spoken briefly to the two women, who were swimming in the pool.

Noah and the White Prince had come in at midnight, at which time the women were plainly visible in their own living room, still in bathing suits, smoking. Valentine, returning at half past one, had noted the living room lights still on, but had not seen anyone inside. Although all their bedrooms had windows that overlooked the courtyard, no one had heard a fight, or the noise of Margaret leaving the house with her bags, or the splash of Ann going into the water, or her drowning thrashes.

Ann was fished from the pool by two ambulance attendants and transported unceremoniously away in a gray canvas body bag, to the delighted horror of the twenty-nine persons gathered in Kiley Court. The police, leaving a few minutes later, took with them a towel found draped on a chair, which belonged to none of the survivors and was assumed to be the property of the dead woman; two clear-glass goblets, one empty and the other nearly filled with an unpleasant-looking purplish liquid which Clarisse conjectured to be an abominably sweet wine; and the minuscule roaches of three marijuana cigarettes found in an ashtray in the living room. Noah was asked not to rent out the apartment until the police had returned and made a thorough investigation.

"Well," said the Prince, who was standing nearby, "she may be dead, but she's still got the place till Saturday noon."

Clarisse, at the gate, detained the last of the policemen. He was young and smug. "What do you think happened?" she asked.

"Accidental drowning. Too much wine and too much grass and not much judgment."

"She was a good swimmer," said Clarisse.

He shrugged. "All right then. It wasn't accidental. It was suicide. Satisfied?"

"No!"

"Look, what do you want? She and her friend that ran off, they were lesbians, weren't they?"

Clarisse nodded.

"Then what happened is: they got drunk, they got stoned, they got in a fight, one skipped out and the other slid under the water."

"Well," said Clarisse after a moment, "I've seen it on the late show."

"Once every summer," said the policeman with a smirk, "but usually not so early in the season. Usually August. Once they even waited until after Labor Day. When we find this other woman, we'll know for sure. The dead girl was pretty small—she must have been the femme."

"The *what?*"

"You know: the wife. The feminine one. What was the other girl like? Built like an eighteen-wheel diesel, I'll bet. They all go in pairs like that, you know. One femme and one butch."

"Have you ever tried writing for *Midnight?*" asked Clarisse. "I'm sure they would find your *aperçus* of gay lifestyles quite interesting."

He shrugged again and glanced back at the pool. "It's classic. I've seen it at least fifteen times. Listen, tell your landlord he better drain and clean that pool before anybody uses it again."

"Aren't you going to dust the surface of the water for prints?" asked Clarisse coldly, and latched the gate behind him. She went into Noah's living room where she found Noah and Valentine drinking coffee and exchanging glances of distress. The two men, hastily aroused from their sleep and confronted with death, both looked the worse for wear. Their appearance did not fit at all well with the simple, even cold elegance of Noah's living room.

Clarisse seated herself at the opposite end of the sofa from Valentine, and took the coffee that Noah poured for her.

She related the policeman's theory.

"Ann Richardson didn't strike me as a potential suicide," remarked Noah. "She was more the victim sort I think."

"Why do you say she was a victim?" asked Valentine.

"I don't know. I suppose that I look on anybody who drinks that much liquor and smokes that much dope as a kind of victim. Something always seems to happen to people like that, probably because they're never really on top of things. I saw Ann outside at the pool yesterday morning at nine o'clock—and she was smoking grass. At eleven, she was guzzling down screwdrivers."

"Oh," said Valentine, "then you mean *accident* victim, not *crime* victim."

Noah nodded. "This wasn't murder, was it?"

"Where's Margaret?" demanded Clarisse.

Valentine sighed in exasperation. "It's too early to play *Clue*, Lovelace. That cop may have had an unenlightened attitude, but he may just be right. Remember Ann and Margaret weren't real lovers, they were only P'town lovers—they both had women back home. Maybe Ann had too much grass and too much liquor and too much romance, and tried to force Margaret into making a commitment. Margaret said no, Ann said get out. Margaret got out, and Ann started to look for coins at the bottom of the pool."

"What if it was murder?" asked Clarisse.

"A swimming pool surrounded on three sides by an inhabited house is hardly ideal as a scene for a crime," said Valentine. "If it had been a deliberate drowning there would have been a commotion, but nobody heard any splashing or anything else last night. There would have been at least one scream, but we didn't hear that either."

"We don't even know if she drowned!" cried Clarisse.

"You're impossible in the morning," said Valentine.

"I mean, don't you two think it's a bit odd that two people have died in this town since Saturday night, both of them gay, both of them found in the water, and both presented themselves first for *my* inspection?"

"And both of them had only one vowel in their first names," said Valentine. "Make something of that."

Though her inclination was to return to bed for the remainder of the day, meditating on the insecurity of human life, Clarisse was advised by both Valentine and Noah to go into work. She did so, and for the first time was actually pleased by the number of customers who, for short periods of time, kept her from conjuring up visions of Ann's face in the water. When Valentine brought her lunch, she shut the door and turned the OPEN sign around to read CLOSED. They pushed aside some fishermen and clowns, and spread the plates and sandwiches and drinks.

"I was sharp with you this morning," apologized Valentine.

"Yes, you were," said Clarisse, and no more was said. "Tell me, how did you end up with Axel last night? We saw them go off arm in arm."

"Unfortunately, Mount St. DeVoto erupted again at the

door of the A-House. Axel left, and he was the first person I saw when you left me at Back Street. He was in sore need of consolation, liquor, and cuddling. I saw to it that he got all three."

"I didn't know he had come home with you. No wall-banging to announce the presence of a second party in your room."

"We fell right asleep. All he needed by that point was a couple of arms around him. You know, you could probably use a dose of consolation, liquor, and cuddling too."

"I'm getting it," said Clarisse. "Tonight. It's my date with Matteo."

"Can I give you a piece of advice?"

"If you do, I pour the Perrier over your head."

Valentine shrugged. "I'll risk anything for your welfare. If you want to snag that cop for a summer fling, don't talk shop. Don't ask about the progress of the investigations. Don't ask him if they've located Margaret of Toronto. Don't ask him if he can sneak you in for a second gander at Ann's body."

Clarisse poured the water over Valentine's head.

But she took his advice. That evening Clarisse and Matteo had dinner at Ciro and Sal's, and then drove to nearby Truro. They sat and talked on the edge of the sand cliffs, and then returned to Kiley Court where Clarisse expressed a desire to show Matteo how her percale sheets worked. Not once that evening did either of them mention the murder of Jeff King, or Ann's suicide. And it was only when Matteo lay sleeping with his head pressed against her breast and the room was quite dark that Clarisse thought of the naked man she had seen the night before in the courtyard; the man who had dressed in the alley, and who might not have been a dream after all.

Chapter 20

A week to the day after Ann Richardson's death, Valentine lay on an oversized emerald-green beach towel on the slanting strip of sand between the Boatslip and the Bay. Like all the other men on towels around him, he had not got enough sleep the night before, and now—despite the din of a dozen portable radios, each tuned to a different station—dozed under the late morning sun, trying at once for recuperation and a tan. The man on the next towel, getting up, inadvertently kicked sand in Daniel's face. Valentine rose groggily on his elbows and stared drowsily out over the sun-speckled water. Beside him were a bottle of Bain de Soleil, half a pack of Luckys, and the paperback edition, generously smudged with oil, of *No Orchids for Miss Blandish*.

He yawned, looked about him to see whether the configuration of sunbathers had much altered itself (it hadn't), then lighted a cigarette. He buried the match in the sand. He yawned again, and guessed the time to be about quarter past twelve. Craning to see the watch on the arm of the man behind him, he found himself correct within two minutes.

The music at the Boatslip, begun at a low volume half an hour before, had edged louder. It did something to cover the cacophony of the radios. Holding his head back with the cigarette in his mouth pointing straight up into the air, Valentine squeezed lotion from the bottle and rubbed it over his chest and arms. The unpleasantness of that sensation cleared his mind.

He exchanged a polite smile with a man sitting four towels down in the grid of bathers, and wondered if he knew him (no), if he wanted to know him (possibly), and if he would be worth the trouble of polite conversation, assignation, and follow-up (probably not). He turned back and spread more lotion on his legs. Clarisse, he considered, overestimated his libido.

He lay on his stomach and had just shut his eyes again when above the radios and the dull laughter—and even duller conversations—floated the distant but unmistakable yodel of Angel Smith. Valentine raised his head, but did not see her on the deck of the Boatslip. He craned to the left and the right, but could not locate her among the sunbathers.

The yodel came again, this time distinctly from behind him—out over the water. He rolled over on his back and sat up. Several hundred feet from shore Angel Smith stood astride an orange surfboard, gripping with one hand the slender mast that had been attached to it, and with the other deftly maneuvering a cord controlling the purple sail. Swimmers shrieked curses at her when she came perilously close to decapitating them, but Angel yodeled merrily. She glided into a wide arc, neatly avoiding the merging wakes of two speedboats, and turned for shore. Valentine waved, and she waved back.

She beached the craft, flipped it over on its side, and pounded up the beach toward the Boatslip. The swimmers were no longer in danger, but now the sunbathers were. She left deep footprints in the hard-packed sand. Valentine waved again, and Angel broke into a lumbering run across the map of cowering sun worshipers.

Breathlessly, she dropped down on one end of Valentine's towel, excavating a large bowl in the sand beneath. "Oh," she gasped, "what are you doing here? I'm so glad to see you! Where's Clarisse? Give me a cigarette before I die!"

Valentine lighted cigarettes for them both. Angel took a long drag and fell back on one elbow. Her bathing suit was black with a pleated flounce about her hips. She sighed, and a cloud of smoke wafted slowly from her mouth.

"I feel just like Diana Dors in *The Unholy Wife*," she breathed.

Valentine rolled his eyes. "You've been to the movies."

"Every showing this week. Along with the rest of the town. Haven't you been to any?"

"I can only take Miss Dors in small doses."

"Honey, Diana's like me—she doesn't come that way. Anyway, I *have* seen Clarisse there. When I'm coming out, she's going in. And she always has the same hunk in tow. A man's man—if you know what I mean."

"Yes," said Valentine, "Clarisse has been occupied lately."

"Ohhhhh," said Angel, and patted Valentine's thigh. "Feeling neglected?"

"No. Jealous. Matteo is the hottest man in town. And Clarisse got him."

"Matteo is straight, Daniel. You weren't even in the running."

"You know him?"

Angel shrugged. "You think I wouldn't remember a man who looks like that? And when he's a cop in uniform? Matteo's great at stopping fights. Two gay men are going at it, and Matteo comes up and they drop their teeth just looking at him. Straight man and woman are fighting, and he comes up—the woman just melts."

"What if it's two lesbians fighting?"

"He knows karate," replied Angel. "Besides, aren't you *glad* Clarisse found somebody here? You wouldn't want her completely dependent on you for companionship, would you?"

"I don't know."

Angel eyed him closely, pressed her cigarette into the sand, and reached for another. "You *are* jealous," she said, "but not of her. You're jealous of *him!*"

"No I'm not," protested Valentine. "I'm just not entirely comfortable around him."

"Why? Because you think he's going to take Clarisse away from you?"

Valentine shrugged. "Maybe it's just that he's a cop. I guess I've never gotten over the feeling that cops are on the other side."

"I begin to smell the double standard somewhere around here," said Angel with a wagging finger.

Valentine sighed. "Maybe you're right. Clarisse has had

breakfast with enough of *my* tricks, and she doesn't pass judgment."

"You mean, she doesn't pass judgment as to number and frequency. But you can't really be worried about her," said Angel.

"No, I'm not, of course—not really. It's just strange to hear all those noises in her bedroom at night."

"Are you hungry?" asked Angel.

"No," said Valentine, "but I could go for a drink."

Angel considered this, biting the end of one of her braids. "To tell the truth, I was planning to go back to the restaurant and shoot a can of cheese spread down my throat, but I guess I could settle for a gin and tonic."

They got a table just beneath the edge of the deck canopy, so Valentine could remain in the sun while Angel got the shade. Valentine took a sip of his scotch and grimaced—it was thinner than he'd ever think of mixing a drink. He and Angel said nothing for a few moments, and he dropped his head over the back of the chair, trying to allow the sun to work on the underside of his chin. When a few minutes later he raised himself, he found Angel staring blankly at the swimming pool.

"I'd say a penny for your thoughts, but this one might deplete my account. What's wrong?"

She shrugged, and picked at the straining stitching in the canvas of her chair. "I was thinking of poor Ann drowning in your pool last week."

"You knew her?"

"Oh, yes—I introduced her to Margaret, in fact."

Valentine looked at her with some interest, shading his eyes with his hands. *"You?"*

"Yep. Just a couple of days before they came down on the ferry. One night Ann came into the Swiss Miss in Brookline and she was sitting by herself and didn't seem happy to be alone. Not long after that I spotted Margaret waiting to be seated and they caught each other's eyes and were staring craters into each other. I know what that look means so I simply showed Margaret to Ann's table." Angel paused to heave a sigh. "How did I know Ann was suicidal? Now I feel like Cupid with a flamethrower."

"Clarisse doesn't think it was suicide. Clarisse says that people commit suicide in the winter, not in the summer."

"Oh, no," said Angel. "More people commit suicide on July Fourth than on any other day of the year. Sales of razor blades and sleeping pills skyrocket June through August. But God, I miss both of them!"

"You don't happen to know Margaret's last name do you? Or maybe you've got her address?"

"I think her last name is Richardson."

"No, that was *Ann's* last name."

"Then I don't know. But you might ask Ann's boss—he would know. He's got that kind of mind."

"Terry O'Sullivan? How do you know him?"

"The night Ann and Margaret got in, they all came to the restaurant. They had just seen me at the Swiss Miss in Brookline, and were surprised to find me here. I did the whole routine for them. I yodeled my eyelashes loose. Then Ann introduced her boss, and I thought she said his name was Mary instead of Terry, and that set me off again. I couldn't get control of myself until the Prince came up and said I was giving everybody in the place indigestion. Meanwhile, Mary O'Sullivan was sitting there, not laughing."

"Thin skin."

"Well, he was such a droopy, pushy, opinionated little wimp—" Angel broke off suddenly and looked at Valentine with mild apprehension. "I haven't just insulted a friend of yours, have I? Ann said he was renting one of Noah's apartments."

"I know him," said Valentine blandly. "That's all."

"Good. You know, Ann and Margaret invited me over to Noah's to use the pool, but I figure I see enough of the White Prince every day without going to his house, so I asked them to meet me here. So on Monday and Tuesday morning last week we had our own little pool parties. They were lots of fun, but Mary O'Sullivan was always hanging around, talking shop with Ann. He *hated* Margaret—probably because he knew Ann's *real* lover back in Boston—and he treated me like the only book I had ever read was the one published by Ma Bell. Ann tried to get it across to him that she was on vacation, but he wouldn't take a hint."

"Thick head."

"Ann didn't want to say anything obvious—he was her *boss*, after all—but Margaret and I got her courage up. She went to his room here at the 'Slip but when she came back she was upset. Apparently he 'threatened her job security,' the little twerp. Margaret wanted to rip his face off, but I told them that they'd have to avoid the Boatslip and that was the end of our pool parties. Next morning, Ann was dead. I'll bet Mary O'Sullivan rides herd over his secretaries with a bull-whip and a cattle prod." She picked up her glass and drained half of it. She coughed. "They always make my drinks so strong here," she wheezed. "It's an odd thing—people see me once, and they don't forget. Sometimes it doesn't pay to be so memorable."

"No," said Valentine, "I suppose not."

Angel sat up with a start. "What time is it?" she cried.

Valentine grabbed the wrist of a blond-bearded man in a transparent white bathing suit just then passing, and studied his watch. "Little after one," he said, smiling at the bearded man, who seated himself at the next table.

"Movies start at one-thirty," cried Angel. "Gotta flee. Don't you want to go?"

Valentine glanced at the next table. The bearded man subtly shook his head. "No," said Valentine.

"Can't I convince you? They're doing a special matinee of *An Alligator Named Daisy* and *I Married a Woman*—Diana's Hollywood period."

"You go on," said Valentine. "I just formed myself a previous engagement."

Chapter 21

"Ahhhhhhhhh," sighed Valentine in guarded ecstasy, but then he pulled his breath in sharply, "Oh, God. Oh, God, damn!"

He had lowered himself into the tub of cool water and now carefully reclined his head against the tile wall. He gingerly laid his arms on the edges of the cold porcelain.

"I'm miserable," he said.

"You know," said Clarisse, standing over him with two ice trays, "this is the first time I've seen you naked since you bought me that Polaroid for Christmas two years ago." She flung a couple of ice cubes into the water. "I have all seventeen pictures taped to my mirror."

Valentine's legs, stomach, and chest were bright red. The back of his body was pale in comparison. His face, on the right side, wasn't so badly burned, but it was much redder than the left side and gave him quite a Harlequinesque appearance. "I'm so miserable," he repeated.

"I thought once you'd laid down a decent tan you couldn't burn anymore."

"Lies!"

"How long did you sleep?"

"Three hours," he replied. "I had lunch—in someone's room—and then went back out on the beach. I didn't think . . ." He shivered violently, and in hope of relieving him Clarisse tumbled more ice cubes into the water. "Oh,

God, I don't want to talk about it. Do you think I'll peel? I'll look like a leper on holiday. I'll—"

"Hysteria won't help," said Clarisse sternly, and turned the trays upside down. Valentine jerked his legs out of the way of the falling cubes. "I'm doing everything I can." She uncapped a bottle of baby oil and squirted a generous amount onto Valentine's chest. He began gingerly to rub it in, but she commanded, "Leave it there. Let it run down into the water." She took another bottle of pink skin lotion and poured that in too.

Valentine squirmed. "Do you have any idea how unaesthetic this is?"

"Great effect though, those ice cubes coated in pink lotion."

Valentine swept a little armada of them away from him. "I've never heard of this cure for sunburn. I thought you were supposed to use very hot water and baking soda. Are you sure you've got it right?"

"From this month's *Cosmo*. It works on sunburn, dry skin, and depression. Look, now that you're all settled, I'm going to get dressed."

Clarisse was wearing her *Come Back, Little Sheba* drag—a full-length floral silk kimono and fluffy pink mules with wooden heels. "One more thing, though," she said, going out into the hall.

Valentine heard her heels clacking down the stairs, and whispered, "I want to die. I want to die . . ."

Clarisse returned with a tall glass, a quart of Perrier, and a dish of sliced limes. "You have to replenish your precious bodily fluids." Valentine stuck two slices of lime beneath his tongue, threw back his head, and upended the bottle into his mouth. Clarisse frowned, leaned against the sink and critically examined her nails. "Why did you go to the beach anyway? Why didn't you just hang around here?—you never get burned out in the courtyard."

"Because when I went out this morning the Prince was giving himself a major facial."

"The ninety-five-dollar job?"

"He was sitting in the lotus position in the middle of a blanket. He was wearing a turban and a jockstrap Mr.

117

Fredericks would be embarrassed to stock. He had two rows of jars and tubes spread out in front of him, and was using a little out of each one. And he was whistling a medley of Judy Garland hits.''

Clarisse fished a snood from the shelf above the toilet and shoved her hair into it. She pushed away from the sink and stepped through the door into her bedroom. She snapped on the bedside lamp, sat on the edge of her bed, crossed her legs, and began to file her nails. Looking up, she could see Valentine suffering in the bathtub. The last light of day filtered weakly through the window at her side. The blades of the window fan whirred slowly and cast a fluttering shadow across her face.

"It's a lesson," she said.

"A lesson in *what?*"

"Social snobbery. It wouldn't hurt you to sit and chat with the Prince once in a while." She pointed her file at Valentine. "You might learn something."

"Oh, yeah," replied Valentine, holding up his hands and grimacing at the water and oil and pink lotion that dripped from them. "Like why Noah skipped town an hour after Jeff King was killed? Or what Noah and Jeff's relationship *really* was? Or what Noah was *really* doing in Boston? Or—"

"Exactly," said Clarisse. "The White Prince thinks women have the intelligence of doorknobs. He tolerates me, of course, but he would never trust me with any real dirt."

"If you want to know all those things . . ." Valentine trailed off with a smile.

"I do!" cried Clarisse.

"Then ask Noah." Valentine shrugged, and gasped, for the smallest stretching of the burned skin was painful.

"I haven't had time to talk to him lately."

"Too busy consorting with a uniform."

"It hasn't been all play," said Clarisse defensively. "I've been pumping him too."

Valentine snickered, and Clarisse kicked the door shut.

A few moments later she opened the door again, and said, "Don't you want to know what I've gotten out of Matteo?"

"About Jeff King?" asked Valentine, eschewing the obvious sarcastic remark.

"Yes."

"You didn't learn anything. If you had, I'd know it by now. Besides, Officer Montalvo doesn't seem the type to relay office gossip to the curious public."

Clarisse dropped the nail file onto her night table and moved back into the bathroom. She sat on the closed toilet seat, took her Lady Norelco from its case, plugged in the cord, and began to run the shaver over her calves.

"Must you do that in front of me?"

"You're sitting there naked, fifteen shades of red, covered in oil and pink lotion, and you're talking to me about *appearances?*"

Valentine took another swallow from the bottle of Perrier. "So," he said finally, "what has Matteo told you?"

"Nothing we didn't already know."

"I thought as much. The question is, is he holding back?"

Clarisse considered this, then shook her head. "I don't know. I don't think so."

"Did you tell him about finding Jeff King's shirt at Maggie Duck's Duds?"

"No," she replied emphatically. "How do you think that the police would react to your wearing material evidence in the bars? They already have a low enough opinion of this household, when a woman drowns in our pool in the middle of the night and nobody hears anything. Besides," Clarisse went on with a grimace, "I don't think the local constabulary is very upset that one of the most notorious pushers in the community has been recalled."

"How are they doing on locating Margaret?"

"Well, they got the photographs that Ann took at the Garden of Evil party, so they know what she looks like. I suppose they're combing the streets of Toronto looking for a woman dressed as Clara Petacci. When they give that up, Matteo said he'd try to get the pictures that were taken of us."

"A souvenir of a night I don't particularly care to remember."

"Val, don't you think it's strange that *nobody* knew Margaret's last name, and *nobody* had her address?"

"I can think of about seventy-five million men who don't know my last name, address, or telephone number. And I'd like to keep it that way. In this town, what difference does your last name make? There're so many more important questions to ask when you first meet somebody."

"Like?"

"Like, 'Do you get tied up on the first date?' "

"All I can say is," said Clarisse, "is that it was very convenient for Margaret to make her exit the very night her lover died."

"Think, Lovelace. Maybe it was like the policeman said—they had a fight, and Margaret skipped out. She had no idea that Ann would commit suicide."

"But what if it didn't happen that way?"

"What else could have happened?" Valentine went on. "All right, say it was an accidental drowning. Say Margaret walks out the door with a trayful of drinks, and there's Ann floating belly-up like a goldfish—"

"Disgusting!"

"—and Margaret says, 'Oh, my God, she's dead!' She panics, drops the tray, packs, and drives away."

"It couldn't have happened that way—we didn't find any broken glass. Besides, she didn't have a car. They both came here on the ferry."

"So she went over to the pier and waited for the seven-thirty bus."

"Why didn't she call the police?"

"She didn't want to be involved. She'd have to answer all sorts of questions. Ann's lover would come down from Boston and there'd be a bad scene. Her own lover in Toronto might find out about it and there'd be an even worse scene. Who knows what she was thinking?"

"That's irresponsible, though. Margaret didn't seem the irresponsible type."

"No, she didn't," agreed Valentine. "But you've got to remember that this is a resort. The only scruples in town are the ones you pay three ninety-five for at the bookstore."

"I don't know what to think," sighed Clarisse.

Valentine looked down at his chest. "Maybe Ann was badly sunburned, and that's why she committed suicide."

"I think you're overreacting to that burn. Men make terrible patients. They can never see the bright side of physical infirmity." She snapped the shaver off, took out a jar of cream and smoothed it over her legs. "You should get out of that tub now. We have to be at the theater in half an hour."

"I'm not going."

"No arguments. You'll sit for three hours in air-conditioned comfort and never move a muscle. Diana Dors will make you forget your pain. Tonight it's *Lady Godiva Rides Again* and *Confessions of a Driving Instructor*—two minor items of her repertoire."

"Am I ready for this?" groaned Valentine. "Where's the boy in blue tonight? Why can't he take you?"

"I told you. He's on night duty."

"Maybe a dark theater is better. And afterward we'll go to a restaurant where I can turn my burned side to the wall. But listen, you know what I could use right now?"

"What?"

"A couple of joints. A little grass would wipe away a lot of pain. Do you have any?"

"As a matter of fact . . ." She went into her bedroom, rummaged in a drawer, then held up a thick plastic envelope.

"Did you get that from Richard?" asked Valentine, refer-ring to a man who worked in Clarisse's real estate office in Boston.

Clarisse's face clouded. "No. Actually I . . ."

"Where did you get it?" asked Valentine curiously

Clarisse's smile was a little grimace. "From the boy in blue."

"Jesus!" Valentine laughed, and pulled the plug.

"Well," said Clarisse, "he just gave it to me so we'd have it to smoke when he came over here. I was flattered. I mean, this is good stuff—and he can't make *that* much money if he's always moonlighting."

Valentine shook his head. "You don't think he *paid* for that grass, do you?"

Clarisse was puzzled. "Well, how else—"

"Dummy. It was confiscated in a raid. The cops pull in a pound and a half of grass, hold on to a couple of ounces as

121

evidence, and the rest gets divvied out. Same thing happens in Key West."

"Oh," said Clarisse, sheepishly admitting her naiveté.

"Well, as long as you've got it," said Valentine, "roll a couple of joints. Mine always come out looking like party favors." He stood in the empty tub. "God, I am *covered* with pink scum."

Chapter 22

After the Diana Dors films Valentine and Clarisse returned to Kiley Court, and talked idly at the poolside until Valentine's itching and scratching drove Clarisse out of her mind. She took him inside, and while he stood in front of the window fan in his room, she changed the sheets on his bed so that he might sleep more comfortably. After he had climbed into bed, she turned out the light, kissed him good-night, and then went downstairs to find out who murdered *Laura*.

Fresh sheets didn't help Valentine's discomfort very much, and the fan only blew hot air over him. He quietly got up and without turning on the lamp sprinkled talc generously over his body, pulled on light clothing, and sneaked out of the house through the kitchen. It was half past twelve and Valentine made directly for the meat rack in the center of town.

When Valentine reached the courthouse, that vast white wooden Victorian structure from which all directions and all distances are measured in Provincetown, he found the benches before it lined with men waiting patiently for the one A.M. bar closings. He lingered near the curb under a street-lamp smoking and watching no one in particular but catching bits of gossip and rumor about people he had never heard of. Two young women, no more than nineteen Valentine estimated, appeared several feet away from him, exhausting their repertoire of sultry looks and sundry winks in a futile effort to catch his eye. Nor was he the only man who thus withstood their charms.

Valentine crushed his cigarette beneath his sandal and wandered through the crowd checking out the new arrivals until he found himself at the end of the sidewalk that led to the wooden steps of the courthouse. He stopped short when he saw Clarisse sitting on the bench nearest the building at the end of a row of four men in high denim. She had one leg drawn up onto the edge of the bench and was examining a piece of paper with the aid of a tiny flashlight held in one hand. Her face bore an irritated scowl. A bright melon-colored envelope lay crumpled next to her other foot. Valentine ambled over and stood quietly before her until she became aware of his presence and looked up sharply.

"I've been swindled," she growled. "Fraud has been committed and I am the victim."

"What are you doing here?" he asked.

"I couldn't sleep and I want to see Matteo."

"I thought Matteo *patrolled* the meat rack, not *cruised* it."

Clarisse stabbed an outstretched thumb toward the courthouse. "He'll be in the station, but not until one." She thrust the paper up at Valentine, angrily snapping the flashlight off and sliding it into her back pocket. "Look what Beatrice has done to me, and I haven't missed a day of work yet!"

The man next to Clarisse looked quizzically at them both.

"It's all right," Valentine said to him, "she's my ex-wife."

The man got up and Valentine took his place. The oblong of paper was Clarisse's first biweekly paycheck from the Provincetown Crafts Boutique. The amount of the check was even less than what Valentine had given her to understand was to be her salary.

Clarisse flicked a finger at the check. "When I went by the shop, Beatrice was just closing up, so I stopped in to speak. She handed it over, without a blush. I could have made more money hawking needlepoint swastika kits on Miami Beach. How am I supposed to live in Provincetown on *that?*"

"I don't understand why it should be so low," said Valentine, handing back the check. "You should talk to Beatrice about it."

Clarisse grimaced as she looked at her paycheck once more before folding it and slipping it into the pocket with the flashlight. She kicked the crushed envelope with the toe of her shoe and then bent to retrieve it. She unwadded the envelope and poked inside—and discovered a slip of paper she had not seen before. She read it, groaned, and handed it to Valentine.

On the page, neatly typed, was a more or less accurate record of all the items that Clarisse had destroyed during the period she had worked in the shop so far, their wholesale price, and the total of the breakage. This sum, along with federal withholding tax, Social Security, state withholding tax, unemployment insurance, and Blue Cross/Blue Shield, had been deducted from her check. Beatrice had appended a little note in her crabbed script that Clarisse was of course responsible for "all the little accidents" within the shop.

Valentine laughed, then caught himself when he saw Clarisse's irate expression. He sheepishly handed the paper back and she ripped it into shreds. He lighted cigarettes for them both and they smoked in silence.

When Clarisse had calmed down a bit she asked, "What are *you* doing out, by the way? I thought I had tucked you away for a long midsummer night's dreaming."

"I'm looking for a hot man with an air conditioner." He craned his neck. "The second shift is arriving so I'd better get to work." He stood, but lingered before the bench. "I hate myself for even thinking this, Lovelace, but are you falling in love with your cop? I've never seen you wait on a bench for anybody. I'm not sure I've ever even seen you on time before."

"No," she said seriously, "I'm not in love, but I do have a good reason to want to see Matteo now."

"Level with me, but make it quick. I'm starting to itch again."

"Don't get angry, promise?" She leaned forward earnestly. "I was sitting downstairs reading and suddenly it hit me."

"What?"

"That Ann's death was . . . was a sort of *diversionary* tactic. To take people's minds off Jeff King." A clock down the street struck one, and she sat back heavily. "You think I'm a fool, don't you?"

Valentine scuffed one heel against the pavement. "No," he said slowly. "I think you're out of your mind. What do you mean, 'To take people's minds off Jeff King'? You're the only one in town who even remembers that man's name. That was ages ago. We're into our second generation of tourists since the Garden of Evil party." He scuffed his heel again, and wagged his head from side to side. "But give me your report in the morning anyway."

When Valentine had moved on, Clarisse left the bench and went around the side of the courthouse to the police station.

Chapter 23

The following morning Clarisse sat at the kitchen table on Kiley Court, removing pastries from a waxed bag. She had gone to the bakery in exchange for Matteo's having made coffee for them. Clarisse wore a white sundress scatter-shot with tiny semaphore flags. Matteo wore a pair of white painter's pants and a pale blue air force shirt with epaulettes.

He was raising the pot to pour their first cups when Valentine's tread was heard on the stairs. An unfamiliar step clattered along behind him.

"This is my favorite part," said Clarisse.

"What?" asked Matteo.

"Seeing who Valentine picked up the night before. Let's make bets. I bet five dollars he's not over five six, I bet he's got a mustache, I bet he's got short hair—curly but probably permed—and I bet he's got a first name that begins with a vowel."

Matteo brought the coffee to the table and seated himself. Valentine appeared, wearing gym shorts, and a recent coating of talcum powder over his burn. Just behind him was a short, mustached, curly-haired man.

Valentine's eyes were heavy-lidded. "Clarisse," he said perfunctorily, "this is my friend Russ. Russ, Clarisse. Matteo, this is my friend Russ. Russ, Matteo."

"Get out your wallet," said Clarisse to Matteo.

"You were wrong about the vowel," Matteo pointed out.

Valentine pointed out a chair for Russ. Russ seated himself with a little *ouch* and a grimace.

Clarisse smiled broadly at him.

"Do you live in Provincetown?" she asked.

"I'm from Providence, Rhode Island," he said. He wore tangerine drawstring pants and a green cotton shirt cut like that of a hospital orderly. He stuck his hand down the back of his pants and rubbed.

Valentine wandered blearily around the kitchen. "What am I looking for?" he asked no one in particular.

"Two cups," suggested Clarisse. He went to get them.

Russ continued to rub himself vigorously. Matteo looked at him askance, and sipped at his coffee.

"Did you hurt yourself?" asked Clarisse, with another smile.

Russ cast a glance of annoyance at Valentine. "I don't know *what* I'm supposed to tell Joe."

"Who's Joe?" asked Clarisse.

Valentine, having poured the coffee, brought the cups to the table. "Joe is Russ's lover," explained Valentine.

"I don't know *what* I'm going to tell him," Russ repeated petulantly.

"About what?" asked Clarisse curiously.

"About *this*." Russ stood up, turned around, and quickly lowered his drawstring pants. On his left buttock was a long purple bruise, shot through with broken capillaries.

Matteo scraped his chair around and faced away from Russ and his injury toward the open door of the kitchen.

"My," said Clarisse mildly, "how did that happen?"

Valentine smiled.

"He got carried away," Russ said accusingly.

"Last night hardly falls into the category of 'carried away.'"

"Well," said Russ, pulling up his pants and sitting down, "what *am* I supposed to tell Joe."

"Tell him you fell off the back of a fire truck," said Clarisse.

Russ looked at her blankly. "This is really serious," he said after a moment. "Joe is going to think I came to Provincetown and got involved in S and M."

Valentine rolled his eyes. Matteo shook his head and got up to pour himself another cup of coffee.

Clarisse smiled more broadly still.

"Try pancake—number five," she suggested.

"I shouldn't have come here last night," Russ moaned to Clarisse, whom he thought to be sympathetic because of her smile. "I should have driven straight back to Providence. I called Joe and told him I was staying with my aunt in Truro. I'm going to go back and he's going to see that bruise, and he's going to ask me if my aunt hit me or something."

"If I hear another word about that bruise . . ." said Valentine, not entirely under his breath.

"Well, you did it!" cried Russ, turning sharply on Valentine.

Valentine remarked to Matteo, "People with tender skin who don't want to get bruises ought not to scream out, 'Oh, God! Oh, God! Hit me, hit me!' when they're just about to"—he paused significantly—"have a good time."

"That was just a figure of speech," Russ protested.

"I'm late for work," said Matteo. "See you later, Daniel. Hope you heal, Russ." He winked at Clarisse and walked out of the kitchen.

"Wait for me," cried Clarisse. Matteo stopped at the corner of the pool. Clarisse ran upstairs for her bag, and when she came back through the kitchen she said soberly, "Call me later, Val, it's important." Turning to Russ, she advised, "Give Joe my best, and remember—pancake number five."

Chapter 24

That noon, when Valentine opened the doors of the Throne and Scepter, he saw Clarisse across the way signaling him violently through the plate glass window of the Provincetown Crafts Boutique. He waved, smiled, and went back inside. He dialed her number, and by stretching the cord, he could see her pick up the telephone across the way.

"Why didn't you come by here?" she demanded.

"It's taken me this long to get rid of Russ."

"What's happened to your judgment? I've come to expect better of you."

"The sunburn went to my brain. I wasn't responsible. So what's up? What couldn't you talk to me about in front of Russ?"

"Not Russ. Matteo. Last night I asked him to let me see the autopsy reports on Jeff King and Ann Richardson."

Valentine laughed. "What did Matteo say?"

"He said no. Then I said, 'It's customary for the person who found the body to initial the medical examiner's report.'"

"And he said?"

"He said, 'You must think I'm a real idiot.'"

"And you said?"

"I said, 'Matteo, if you let me look at those reports, I'll . . . I'll *go to bed with you!*'"

"Like you have for the past six nights and last Monday afternoon?"

"Don't belittle a woman who has just traded her body for vital information."

"What *was* the information?"

"Most of it was technical stuff, and measurements in centimeters—"

"That could be interesting."

"But the main thing was that Jeff King was stocking a drugstore in his stomach. All sorts of things."

"But it wasn't the drugs that killed him, was it?"

"No, and it wasn't the bump on his head either. It was definitely strangulation. The time of death was placed between two and ten A.M."

"What!"

"That's what the medical examiner has in the report, despite the fact that the police have my sworn testimony pinning the time down to between three-thirty when I saw him jump into the bay and a quarter to five when I found him on the beach."

"I guess the coroner likes to give himself a little leeway."

"I'd prefer accuracy. Makes you wonder about the rest of the report."

"What about Ann?"

"That was drugs. Angel dust and MDA—and lots of it. She mixed it with the wine and drank it."

"Not much doubt of suicide there. What's that phrase, Lovelace? *I told you so.*"

"We can't be sure until Margaret shows up."

"If we knew her last name . . ."

"*Nobody* knows her last name."

"Do you think your uncle might be holding back?"

"No! If Noah knew, don't you think he'd tell? I'm sure he doesn't like the idea of tenants taking death-dives in his pool."

"How does he feel about ex-lovers taking death-swims in the bay? When are you going to sit down and clear the air a little?"

"As a matter of fact, Mr. Valentine, I'm taking my lunch hour at Kiley Court this afternoon. I will seduce Noah into intimacy over cold avacado salad."

"You never fix avocado salad for me."

"But I do make polite conversation with your battered boyfriends. . . ."

When she got to Kiley Court, Clarisse skirted the empty swimming pool, and opened the screen door of Noah's apartment. The interior was quiet and cool. The noon sunlight, filtered through the vines that covered the windows, spangled the newly laid hall carpet. She padded down the passage toward the kitchen but stopped short before the double doors to the front parlor. In an oval mirror on the unwindowed wall of the room she caught the image of Noah sitting, with his back to her, at his desk. He apparently had not heard her enter.

His posture—tense on the edge of the chair—was so uncharacteristic of Noah's usual physical ease that Clarisse remained as she was, observing. Noah suddenly raised his head and glanced into the mirror. Clarisse snapped back out of the way. She dared not look into it again—knowing that if she could see him, he could also see her.

Clarisse realized, with a mixture of amusement and distress, that she was spying on her uncle. She briefly considered the morality and the embarrassing possibility of discovery— then once more peered into the mirror.

Noah sat back suddenly on the chair, and the alteration of his position allowed Clarisse to see that he had been reading a letter. He folded the single typed sheet, and slipped it into a fresh white envelope, the flap of which he licked and sealed. The envelope in which the letter had arrived lay atop the desk. He ripped this in two, and slid the pieces into his back pocket. He tapped the plain white envelope twice with his forefinger, and then lifted the blotter and placed it beneath. He carefully realigned the blotter, ran his hand over it to see whether the bluk of the envelope could be felt beneath, then stood and walked directly toward the hall.

Clarisse panicked. She turned on her heel, slapped her hand against the screen door so that it opened and slammed with a bang, dropped to her knees and frantically patted the carpet with her outstretched hands.

"Clarisse!"

She tossed her hair back and looked up over her shoulder at her uncle standing behind her. She squinted one eye.

"Oh, hello," she said and went back to patting the carpet.

"What are you doing here? Oh, that's right—we're supposed to have lunch . . . did you lose a contact?"

"I tripped on the doorstep and it popped out. I know it's right here somewhere."

"Don't let me step on it," Noah said, retreating carefully into the living room. Clarisse continued to pat the surface of the hall runner. Noah said, "Listen, do you mind if we put off today? Angel needs me at the restaurant. High-level decision making."

"No," said Clarisse looking up with a fake squint. "Of course not."

"Well, then I'd better go upstairs and get ready. I'll leave you to this. I'd help, but I'm afraid . . ."

She waved him away. "I'm used to it."

Noah went up the stairs. As Clarisse looked after him, she could see the outline of the torn envelope in his pocket. She listened to his footfalls' progress down the upstairs hallway.

Clarisse knew that she should leave the house; instead she got to her feet, slipped out of her sandals, and crept into the parlor on the balls of her feet. She took out the envelope and with a letter opener unsealed the still-damp flap. Inside was a single-page letter from the office of Calvin Lark, the lawyer in Boston who represented not only the rental agency where Clarisse had worked but her uncle as well. The letter of intent was addressed to Noah Lovelace and explained in para-legal terms that, following Noah's instructions, Calvin Lark would alter the names of the beneficiaries of his will and several insurance policies. The name of Jeffrey Martin King would be excised from all documents, to be replaced with that of Clarisse Lovelace. She slipped the letter into a fresh envelope, sealed the flap, and shoved it back under the blotter.

She softly left the house, crossed the courtyard, and entered her own kitchen. She made some fresh coffee, prepared herself a salad, and began to reflect on the revelation that she had taken the place of Jeff King as her uncle's principal heir.

Chapter 25

Clarisse was late in opening up the Provincetown Crafts Boutique. She stood before the door and fumbled in her bag for the keys. The door swung open to reveal Beatrice standing there smiling.

Clarisse looked up, took one short breath, and hastily explained, "My baby brother had this terrible accident—he's twenty-one but I still call him 'baby brother'—and I had to give blood."

"I was walking by and I saw the shop wasn't open," said Beatrice. "I looked at my watch, and I said, 'Oh, my, it's already a quarter past ten and where is Clarisse?' I hope your brother is going to be all right. Did he have an accident?"

"Run down by a fiend on a skateboard, right in front of the Massachusetts Statehouse."

Beatrice shook her head in sympathy. "But you know," she said, as Clarisse swept past her into the shop, "if you're going to be late like this, you should give me a call. I don't mind opening up when there's a real emergency."

"I'm sure it won't happen again," Clarisse smiled. She settled herself behind the counter, and plugged in the register.

"Shall I flip the sign?" said Beatrice. Clarisse sighed and nodded.

No sooner had the placard been changed from CLOSED to

OPEN, than three pre-teenaged girls entered and demanded use of the bathroom.

"I'm afraid our facilities are not for the use of the public," smiled Clarisse, and glanced at Beatrice.

"I *got* to go," wailed the smallest of the girls.

"Then don't stand near my display cases!" cried Beatrice, and hustled the girls out. She came back in and said, "Clarisse, I'm expecting a shipment from Chicago this morning. Five dozen throw pillows in the shape of oysters. See if you can find a barrel or something in the back we can display them in. Don't you think that would be cute—a barrelful of cloth oysters?"

"Darling!" cried Clarisse vehemently. "Precious! They'll probably all be sold the first day!"

Beatrice hesitated in the door of the shop, then turned and came close to the counter. "Let me ask you something—as long as there's nobody else around."

"Yes?"

"How do you like working here?"

Clarisse glanced around the shop. "I'm overwhelmed," she said after a moment.

"Good," said Beatrice. "I wasn't sure at first. There was so much breakage, I thought that perhaps you were subconsciously expressing your unhappiness by destroying my merchandise. I remember when I used to work on my grandmother's chicken farm every summer, I'd break half the eggs getting them out of the nests."

"I'm more used to the place now," said Clarisse.

"That's good," said Beatrice vaguely. "Now, the reason I ask is this: I've been offered the opportunity to open a branch store in Boston, down near the Waterfront–Quincy Market area, and I just wanted to know if you'd like to manage it for me."

Clarisse stammered, "I . . . I'm not sure I know what to say."

"Well, I think you could handle it—you're doing just fine with this place. I'd do all the buying, so you wouldn't have to worry about that. You'd just have to take care of hiring, and making sure the employees show up on time and so forth. I keep this place open all year—I wouldn't miss a Provincetown winter for anything—so I don't want to have to spend two

135

months setting the place up in Boston. Since you're going back in September anyway, I thought you might as well . . ." Beatrice paused significantly.

"Let me talk it over with Val," said Clarisse at last.

"I've already spoken to him," said Beatrice.

"You have?"

"This morning. I ran into him at the Portuguese bakery. He said he thought it was a great idea. He said he had overheard you on the telephone talking to your mother, telling her that this was the best job you had ever had in your entire life and you just wish you could keep it for always."

"I see," said Clarisse.

"Well," said Beatrice with a wide smile, "here's your chance."

A pair of middle-aged female twins in identical yellow pantsuits walked through the door, and Beatrice took her leave.

Clarisse dialed the number at home, and the telephone rang fifteen times—Valentine wasn't there. She called the Throne and Scepter, and the day manager took a message that Valentine was to call as soon as he got in.

Chapter 26

As soon as she saw the doors of the Throne and Scepter open for business that day, Clarisse exclaimed loudly to the half dozen customers she had in the shop that she thought she smelled smoke. When they had hurried out, fearful of their safety, Clarisse turned the Open sign to Closed and ran across the way. Valentine was just plugging in his cash register.

"I could kill you for what you told Beatrice," she said evenly.

"I had to tell her something."

"Now she wants me to manage a branch store in Boston. Can you imagine? The Boston Guardians of Taste will throw bricks through the window."

"Then turn down the offer. Politely. Tell her you're going to be too busy with your first year at Portia Law."

"I *will* be busy."

"The perfect excuse, see? So now why are you so upset? You get to keep the job for the summer, and after Labor Day you can tell Beatrice what you think of her merchandise."

"I'd never do that," whispered Clarisse. "Anyway, I'm upset about something else."

"What?"

Clarisse told him what she had discovered that morning in Noah's desk.

"So," said Valentine, "you're an heiress now."

"That's not the point," replied Clarisse.

"What am I missing? There's something else to Noah's changing the will then?" He shoved a congratulatory drink at her across the bar.

"Haven't you made the obvious connection yet?"

"What obvious connection?" asked Valentine.

"Remember the story Angel Smith told us about Jeff King, and the man who lived below her?"

"Yes."

"Well, that was Noah."

"How do you figure *that?*"

"Remember? Angel said that Jeff King stole six place settings of Rosenthal china from the man who lived downstairs from her."

"It was eight. But so what?"

"So that struck a bell. The Lovelaces have never bought anything but Rosenthal."

"So do a few others."

"But Angel was hedgey about telling us the man's name."

"She figured we didn't know him anyway, so what was the point of giving his name—or maybe she had forgotten it," argued Valentine.

Clarisse ignored him. "So then I called my brother, and asked him where Noah was living eight years ago. He was living on Queensbury Street," she concluded triumphantly.

"So why didn't Angel just *say* that she used to live upstairs from Noah and Jeff King?"

"Angel and Noah are partners. She wouldn't gossip about him."

"All right," said Valentine. "I'm convinced. And it fills in a few gaps."

"But what worries me," said Clarisse, "is *when* Noah changed the will. He went to Boston the morning that Jeff King was killed—and he left even before I got back from the police station. Word wasn't out yet. And Cal's letter says that that was the day that Noah made the changes."

Valentine shrugged. "So talk to him. That's all it takes to clear everything up."

"I can't say anything about that, Val. I'm not supposed to know about the will."

"You don't really think Noah killed Jeff King, do you?"

"No," said Clarisse. "But I'll bet he knows something about it we don't know. There was something else in the will too."

"What?"

"A special bequest of ten thousand dollars for Victor Leach."

"Who's Victor Leach?"

"The White Prince, of course," said Clarisse. "Is it any wonder now we never heard his last name?"

"Ten thousand dollars," Valentine mused.

"That's right," said Clarisse. "Considering the extent of Noah's fortune, ten thousand dollars is nothing more than a polite nod of recognition. It shows us where they stand, doesn't it?"

"I wonder . . ."

"Wonder what?" demanded Clarisse.

"Why Noah waited so long to change beneficiaries. If that will included Jeff King, it had to be at least seven or eight years old. If Noah had died, Jeff King would have gotten everything, and you and the Prince would have been out in the cold. Clarisse, I really think you ought to talk to Noah. There's something strange about all this, and I think you're right. He does know something we don't."

"Who's curious now?"

"It's a slow day, and I need something to keep my mind off my sunburn. Come back when you get off work, and I'll tell you who did it."

"You're impossible. For weeks you've been trying to get me to forget poor little Jeff King, and poor old Ann Richardson, and now that I'm closing in for the kill, you want to share in the credit. Well, Ducky-Lucky, I'm going to do this one all on my own!"

As the afternoon waned business picked up at the Throne and Scepter. Rain had begun to fall only a few minutes after Clarisse returned to the Provincetown Crafts Boutique. It showed no sign of letting up. On Commercial Street Valentine glimpsed a parade of bobbing multicolored umbrellas and flashes of glistening orange and yellow rain slickers. Inside the bar no one had bothered to drop quarters in the jukebox and Valentine had not yet turned on the tape machine. The patter of the rain, the buzz of conversation, and the hum of the ice machine provided a soothing backdrop of noise. The room smelled of wet clothing.

A little after four o'clock, Angel Smith, swathed in a full-length lime green poncho, surged into the Throne and Scepter like a Lake Superior barge. She threw back her hood and swung her arms about inside the poncho, again drenching half a dozen young women who had come in mostly for the purpose of drying themselves off. Angel's hair was braided and she was in full Swiss Miss war paint. Valentine could hear her clogs on the wooden floor.

Lifting up the tent of green canvas she dropped it over a few stools, and hoisted herself up to the bar. The hem of the poncho hovered above the floor, and soon formed a perfect circle of water beneath her.

"What'll you have?" Valentine asked.

"I have to be at work in a little while, so nothing strong. Just give me a vodka on the rocks—in a small glass."

While Valentine was fixing the drink, Angel began to rummage beneath the poncho for money, but this action unleashed so much water that Valentine said: "Don't bother —on the house. You'll owe me one sometime."

Angel took a sip with contentment. "You mix a good drink, Daniel."

"Vodka and ice can be tricky—unless you know exactly what you're doing."

She swallowed the remainder of the drink in one gulp. "Well, no more pussyfooting," she said in a stern voice. "I have come here today as a recruiter."

"For what?" Valentine asked uneasily.

"Next week's show at the Plymouth House cabaret. The man they had hired to MC it got hepatitis from one of the boys in the chorus, and I leapt into the breach."

Valentine paused a moment to consider this image, then said: "Are you going to MC it, or are you taking the place of the chorus boy?"

"I'm going to MC."

"And you want me for the chorus—no?"

"You won't have to say any lines. But it's a great part.

Great costume too—made me hungry just to look at it." She took a breath and closed her eyes in ecstasy. Valentine poured more vodka.

"Get the White Prince."

"He's already signed up. He'll probably want to do that weary Salome number again—coked to the falsies, of course. Unless the cartilage in his nose collapses between now and then, which would be a blessing to us all. I mean," she went on, snatching up a few maraschino cherries from Valentine's side of the bar and popping them into her mouth, "everybody likes cocaine, but not everybody uses it as a substitute for literature, red meat, and love."

"As addictions go," said Valentine, "it's cheaper than swallowing semiprecious stones."

"Be in the show," Angel begged, clasping her hands on the bar between them. "You'd only be on stage for fifteen seconds. No lines—you don't even have to move. You'd just be a mannequin for the costume. No one will recognize you, I promise. I'll do your makeup myself."

"No," he said. "But just out of curiosity—I'm very curious today—what's the role?"

"It's a tableau vivant—Madame Du Barry mounting the scaffold. Oh, Daniel, you should see the dress! Lace at the bosom, and panniers so wide that they're having to stick the executioner down in the orchestra pit."

"What is the show, Angel?" asked Valentine suspiciously.

Angel tipped the ice from her first drink into her mouth, and bit down loudly. "We're calling it *Prostitution through the Ages.*"

Only with two more vodkas did Valentine at last convince Angel that he had no intention of taking the role of Madame Du Barry in the series of tableaux vivants being prepared for the following week. As she was leaving she made the identical proposition to a stranger who was madly drying his recently permed hair with a stack of paper napkins from the bar. Though bewildered, the young man accepted. When Angel had gone, he whispered in awe to Valentine, "I'm from the Midwest. Is Provincetown always like this?"

The rain had grown even heavier and the sky was now quite

gray. The bar was crowded with wet customers and the plants that had been brought in from the patio. Valentine gave a stack of quarters to the waiter, and told him to choose something quiet on the jukebox.

The first selection was Tammy Wynette's "Stand by Your Man."

Valentine wiped up the bar, and rubbed specially hard at the rings left by Angel's several glasses.

"Hey," said the man who had taken her place, "this stool is loose, you ought to have the screws tightened up."

Valentine looked up quickly. The man was Terry O'Sullivan.

Valentine closed his eyes briefly, then opened a bottle of Perrier, jammed in a wedge of lime, and set it before Terry.

"Thanks. You remembered."

"It's part of the job. I get bigger tips if I remember."

"Your skin is so red! How did you get so burned?"

"I met a hot man from San Francisco—he was into sunlamp torture."

Terry frowned. "You're making that up."

"No I'm not." Valentine lighted a cigarette.

After a moment of silence, Terry said, "Have you thought about our relationship anymore?"

Valentine pulled back. He had been waiting for this. "No," he replied.

"I asked you to think about it—and you didn't?"

"No."

"Daniel, I want to tell you something. I see you a whole lot differently than you see yourself. What you need is a stabilizing force in your life. That's what deep down inside you really want. I can see that in everything you do. It's absolutely clear."

"Aren't you late for an est meeting or something?"

"No," said Terry. "The only reason I'm here in Provincetown is because of you."

"I wouldn't have thought you'd come back here."

"*You're* here."

"But your friend died here in Provincetown," said Valentine. "I would have thought you'd be too upset to come back here so soon. I would have been."

"I don't care about Jeff, I care about you, Daniel!"

"Jeff!" cried Valentine in astonishment. "I was talking about Ann Richardson—your administrative assistant, remember, who drowned in our swimming pool? I didn't even know you knew Jeff King."

Terry O'Sullivan suddenly looked very uncomfortable.

Chapter 27

"You knew him, didn't you?" said Valentine.

Terry O'Sullivan turned uneasily on the stool. "Everybody knew Jeff."

"But when I talked about somebody getting killed in P'town, the person who leapt to mind was Jeff King—not the woman you worked with for five years. You must have known Jeff pretty well."

"Yeah, I guess so. No, it's just that Ann wasn't killed, Ann committed suicide. You should see our office now, it's a real wreck. And there's been a budget cut, so I can't hire—"

Valentine got the waiter's attention and motioned him to fill in behind the bar. He mixed a strong gin and tonic for Terry and then pointed him toward a small table in a corner next to the front window. When Terry sat down, his face was framed with palm fronds. Valentine lighted a cigarette.

Terry tasted his drink, made a face, then squeezed out the remaining lime. "You know I don't like hard liquor."

"You'll be thankful for it when we're finished talking. Now what about Ann Richardson? Did she know Jeff King too?" Terry took a swallow of his drink and Valentine read the gesture as a delay tactic. "Ann had some pretty hefty drugs in her when she died," Valentine went on. "Angel dust and MDA. Maybe she got it here in P'town."

Terry remained obstinately silent.

"Did she get it from Jeff?"

"I don't know."

Valentine smiled, reached across the table, and tightly grasped Terry O'Sullivan's shaven jaw. He lifted his face. "Yes you do," he whispered. "And I want to know." He dropped Terry's jaw, and said, in an entirely different tone of voice, "You remember my friend Clarisse?"

"How's she doing?" asked Terry automatically.

"Fine, just fine. She's got a new boyfriend. His name's Matteo. Matteo's the cop in charge of investigating Ann Richardson's death."

"It was suicide. There's nothing to investigate."

Valentine shrugged, as if it were none of his concern to interpret police procedure. "But as long as you're in town for a few days, I'm sure he'd like to talk to you."

"I didn't do anything!" Terry hissed. "It was an *accident!*"

"What was?"

Terry paled and blinked several times.

"Tell me," said Valentine.

"What difference does it make to you?"

Valentine smiled. "I just don't want Clarisse's boyfriend getting on your case, that's all."

Terry O'Sullivan pursed his lips, and swallowed the rest of his gin and tonic. "All right, I sort of knew Jeff King. Satisfied? We weren't close. But he sold grass, and he was always easy to find when I needed it."

"You?" Valentine smiled. "The watchdog of homosexual morality? You smoke grass?" He signaled the waiter, and another drink was brought for Terry.

Terry looked at Valentine sourly. "If you must know, I have . . . I have sexual problems if I'm not a little stoned."

"We've all got problems," said Valentine dismissively. "But why were you buying angel dust and MDA from him?"

"I didn't! You know how I feel about hard drugs!"

Valentine snapped his fingers. "Of course! Why didn't I think of it before?" He smiled sweetly at Terry O'Sullivan. "Jeff King was in *your* room at the Boatslip that afternoon. *You're* the one who put him up, aren't you?"

Terry grimaced with consternation. "Yes. I ran into him at the Boatslip pool Saturday afternoon. He had his bag with him, and he said he didn't have a place to stay and could he stay with me. I said no, and then he said he'd give me an ounce of grass if he could just sleep on my floor. Well, I knew

I could use the grass and I figured I'd end up at your place anyway, so I said all right."

"You didn't have any grass at the party."

"I don't need to smoke at parties, just when I'm going to have sex. So when Ann got sick—Ann always had too much to drink at parties—and her friend and I helped her outside, I ran up to my room and got the grass. I had it with me when we went back to your place. But I never got the chance to use it," he added with a bitter smile.

Valentine thought for a few moments. "And when you left my place that morning, you went back to the Boatslip . . ."

"And I passed the meat rack and everybody was talking about this body that had just been found on the beach."

"Did you know it was Jeff King?"

Terry shook his head. "When I got back to my room, his bag was there, and he wasn't, and I started getting nervous. So I went back down to the meat rack, and when somebody said it was a drug pusher who got killed, I knew it had to be Jeff. Then I got scared."

"So," said Valentine, "you dropped the bag on the doorstep of Maggie Duck's Duds?"

"How'd you know that?"

"Sublime detective work. But you kept Jeff's merchandise?"

"I couldn't walk the streets with a bagful of drugs!"

"So you gave them to Ann instead. Didn't you bother to tell her that you were presenting her with a lethal dose of angel dust? You shouldn't be giving anyone angel dust anyway."

"I don't know anything about drugs," said Terry O'Sullivan with a little shrug of pride. "Jeff told me he was selling grass and Quaaludes and MDA and that's all. So I kept the grass, and I flushed the Quaaludes down the toilet, and then I mixed all the white powder together in one envelope and gave it to Ann, because I knew she liked MDA. Ann would take anything."

Valentine sighed with unhappiness and disbelief. "You mixed drugs without knowing what they were?"

"Jeff told me it was MDA! It all looked alike. How was I supposed to know he was selling angel dust too?"

"Because Jeff King lied with every step he took." Valentine

looked closely at Terry O'Sullivan, who managed at once to look morose and arrogant and greatly put-upon. "Doesn't it bother you that you're the one responsible for Ann's death?"

"It was an accident!" protested Terry. "I'm not responsible. She shouldn't have taken so many drugs. Her friend didn't die, her friend didn't gobble up drugs the way Ann did. You know what Ann used to say in the office? She used to say, 'Show me a drug and I'll introduce it into my system.' And she meant it too!" Terry finished off his second drink. This time *he* signaled for another. "I used to make Ann take her lunch hour at three o'clock. You know why? Because she always came back drunk."

"I thought you didn't believe in drugs—so why did you give them to Ann?"

"Well, Ann and I had had this fight—it was about work— and I came down hard, and I probably shouldn't have. And I wanted to make it up to her. So when I had those drugs, I thought it was just as easy to get rid of them that way as flushing them down the toilet. MDA is so expensive I didn't feel like dumping it."

"Where are you staying?" asked Valentine quickly. He had no interest in Terry's excuses.

"I'm at the Boatslip again. But in a different room."

"Why don't you go back there now? Take a nap or get something to eat. I'll stop by after I get off my shift. There are probably some more things I want to ask you, but I want to think a little first."

"Listen, you're not going to tell that cop, are you? I didn't really do anything wrong, Daniel. I told you, it was an accident."

"It was also very bad judgment, giving away drugs you're not sure of, mixing drugs you're not sure of. You should have gone to the police anyway and told them what you knew about Jeff King."

"They would have thought I did it!"

"You couldn't have done it, you were with me the whole night."

"Please don't tell the police."

"If I did," said Valentine, "they'd probably arrest you for terminal stupidity."

The waiter brought a fourth drink, and Terry finished it off in one long swallow. Without another word to Valentine, he stood and walked unsteadily out of the bar. The rain had lightened to a mist and more people had taken to the street. Valentine turned and looked through the window. The early dusk was gathering and the paling light took on a faint green hue through the rain. Valentine watched Terry O'Sullivan make his staggering way down the sidewalk toward the Boatslip—he evidently *wasn't* used to hard liquor. Valentine sat for a minute longer then rose to return to the bar. When he pushed back his chair, it slammed into another, and he turned to apologize to the person who occupied it.

"Sorry," he said automatically.

"It's all right, Val," replied Noah Lovelace. Across from Noah, the White Prince nodded a curt greeting.

Chapter 28

Standing before the full-length mirror in her bedroom, Clarisse surveyed herself from several angles. She undid one more pearl snap of the starched blue cotton western shirt and, taking a very thin gold chain from a porcelain dish on the bureau, fastened it about her neck. She surveyed each foot, checking for scuffs on her Wellington boots and then slipped a thin leather belt through the loops of her jeans. She peered at the clock on the nightstand, and found that she had three-quarters of an hour before her date with Noah at the Swiss Miss in Exile.

She had conducted a dozen dialogues in her mind, in which she discovered from her uncle his relationship with the recently deceased Jeffrey Martin King, but each experimental conversation had seemed more absurd and rude than the last. She decided to trust to her instinct, and dismissed all further thought of the delicate task before her.

The storm, which had seemed to her to show signs of abating in the late afternoon, had only gathered more force. The rain continued to pelt the panes of the bedroom windows, making them pale silver in the last half hour of daylight. Clarisse lighted a cigarette and leaned against the window frame as she smoked. She stared down into the empty courtyard. Rainwater had gathered almost an inch deep at the bottom of the pool, making it a shimmering black mirror.

149

COBALT

The rain beat down the impatiens in the window box. Clarisse raised the window a little, jumping out of the way of the water that splashed in from the sill, plucked one of the coral flowers, shook the water off it, and stuck it through the top buttonhole of her shirt. She took another drag on the cigarette before grinding it out in an ashtray on the bureau. When she turned back her eyes went immediately to the middle section of the house in the courtyard. The first-floor lights had been turned on and she could see a male figure move past the windows. She wondered vaguely who the new tenant was, but when the man did not appear again she moved back to the mirror. She carefully painted her lips a dark ruby, but decided against any other makeup. She studied herself a long moment, and was dissatisfied. Parting her hair in the middle, she pulled it back close to her head, deftly fashioned a loose chignon, fastening it with delicate combs. She wondered if anyone else would recognize the striking likeness to Vivien Leigh. Before leaving the bedroom she sparingly splashed perfume behind her ears and at the top of her cleavage.

Downstairs in the kitchen Clarisse spooned instant coffee into a mug and poured in boiling water. Provincetown was often too hot for coffee, and she relished the coolness provided by the rain. She sat at the kitchen table, and leafed through Valentine's latest edition of the *Journal of the American Playing Card Association*.

She was two pages into "A Cursory Examination of Cheating at Whist in the Eighteenth Century" when a light rap sounded at the door to the courtyard. She flipped the magazine closed and without getting up, called out, "It's open. . . ."

She turned. Axel Braun stood in the open doorway, dripping wet. He wore only a pair of low-rise swimming briefs. He wiped water from his face and smiled.

Clarisse rose and tore a number of paper towels from the roll and handed them to Axel. "Come on in," she said, "and don't mind the floor."

He set about to dry himself. The muscles of his arms, abdomen, and legs flinched each time he blotted off more water and Clarisse strove to keep her eyes at a respectable level.

"I thought you had left town," she said. "After the unfortunate business of the corpse in the pool."

"Yes," said Axel. "I did. *But,*" he said with a quick nod and a smile, "I'm back now."

"Do you and Scott still have your cottage?"

He squinted at her. "Ah, no," he said, "in fact I'm staying here."

"Here? Daniel didn't say anything—"

"No, not *here,* I mean in the other apartment. I'm renting from Noah for a couple of weeks. Are you getting ready to go out?" he asked, taking note of her outfit. "Don't let me hold you up."

"I have to be somewhere in a while, but I have time to brew you another teaspoonful of instant."

"Thanks," he said, and seated himself at the table. "I'm still dripping," he apologized.

"You're in better condition than most of the men I serve coffee to in here. Where's Scott?"

"New Hampshire. Lake Winnipesaukee."

"I've always believed," remarked Clarisse, "that a good lover was like the international fishing limit—about two hundred miles out. Did he take a wrong turn at the courthouse?"

"He's visiting his sister and her family. But actually it's a trial separation. We're trying to mend things. It's the first time he's ever let me go anywhere alone. I think it's a good sign. He'll be down here in another week or so."

Clarisse allowed herself once again to be impressed by this handsome man's unself-conscious vulnerability. Underneath the rain she could just make out the sound of the courtyard gate scraping open.

"That must be Val," she said, glancing at the clock. "He's due."

Within a few moments, the kitchen door was pulled open, but it wasn't Daniel who filled the frame. Water poured down off the plastic cover of Matteo Montalvo's policeman's hat and splashed on the quarry tiles.

"I was on my rounds and—" he began, but then caught sight of Axel. From where Matteo stood he could not see Axel's briefs, and it looked to him as if the man were sitting naked at the kitchen table.

Matteo looked Axel over, quickly sizing him up as a potential competitor for Clarisse's lustful affections. Clarisse perceived an unmistakable flicker of jealousy in the policeman's eyes. Axel looked Matteo over also, but with a different intention and a different result in mind. Axel was pleased with what he saw.

Clarisse introduced them, employing first names only, and added, "Axel is Noah's new tenant."

Matteo retreated into the darkness and the rain, letting the screen door slam shut. "Well," he said from outside, "I have to be on my way."

"Wait," cried Clarisse. "Wait, Matteo!" He paused but did not open the door again. She could scarcely see his face through the screen. "Can you walk me up to the Swiss Miss? I'm meeting Noah there in fifteen minutes."

Without waiting for his reply she ran for her umbrella. When she came back into the kitchen, she walked over to Axel and said in a low voice, "The cop is taken. He's mine. If you even speak to him again, I will handcuff you to a leper."

Axel folded his arms across his bare chest and smiled. He nodded a farewell to the cop over Clarisse's shoulder.

Leaving Axel to wait for Valentine, Clarisse stepped out the door and opened the umbrella over herself and Matteo. He was silent.

"Despite what you may think," said Clarisse, "Axel had no designs on me. He's a married man and he never fools around. He just came over to borrow a bulb for his slide projector." She slipped her hand beneath Matteo's slicker and rubbed the pistol in his holster. "Oh, God," she sighed, "every woman ought to have an armed escort. . . ."

Chapter 29

Noah Lovelace had an elegance of demeanor that Clarisse was sure had been achieved only through a long life unencumbered with nine-to-five cares. He did not try to be clever, he did not try to be kind, he did not seek to be envied—yet everyone who came near him was eager to be regarded as his friend, confidant, or lover. He was a man who hid neither his faults nor his troubles nor his disappointments—and because he was so free with these intimacies one had the distinct impression that there were many matters that he was keeping entirely secret. Somewhere within him, Clarisse sensed, there was a wall over which no one had ever been allowed to look.

"You look done in," said Clarisse to her uncle, as she sat across from him in one of the private dining rooms of the Swiss Miss. "Bad day?"

"Bad days seem to come once a week in Provincetown," Noah mused. "Mine are usually Friday."

"Why, do you suppose?"

He shrugged. "Habit, I guess. Nothing particular happened today. I told the Prince that I wanted him to leave at the end of the summer, but I've told him that so many times, it's like saying good-morning. This time I'm serious, however."

"I had no idea you had decided to break up."

Noah laughed. "Did you really think we were together?"

"No," she admitted. "Val and I have wondered exactly what the relationship was."

"The relationship is habit, that's all."

Clarisse looked around the room. "And business too, now."

"And business. But it's time, I think, to break the habit. I've told him that when the season's over I want him to find another place to live."

"How did he take it?" asked Clarisse curiously.

"He didn't."

"Didn't what?"

"Didn't take it. He doesn't believe I mean it. Every week I tell him I want him out, I tell him it's over between us, and he ought to be sending out résumés and checking the want ads."

"You're firing him too?"

Noah shook his head. "Angel and I have pretty much decided to close this place in November. We really only get tourist trade, so there's no point in keeping it open all year. We'll open up again in May—but without the Prince. This winter, Angel will concentrate on Brookline and I think I'll spend a few months in Morocco. I like Morocco."

"Didn't Truck-Stop Betty move to Morroco?"

"Yes, he did. And I'll stay with him. He lives in the medina in Rabat. In a house that overlooks the old Portuguese fortress and the Atlantic. You can stay there about three months before you get overwhelmed by its picturesqueness."

"That'll be good for you. It'll help you over the trauma of the breakup."

He smiled. "There won't be any trauma. Not with the Prince. I think the only thing that will convince him I'm serious is if I lock up the house and the restaurant and disappear for three months."

Beyond the rain-streaked window Clarisse could see that although the garden dining area was deserted, the lights there still burned, casting halos of yellow in the wet evening air. Footsteps moved constantly past in the hallway and the low chatter of customers could be heard distantly and not at all unpleasantly. Angel's yodel warbled in the distance. The waiter appeared with the wine and poured some for Noah to taste.

Noah looked up with a smile. "George, I picked out the cellar here. If the wine's no good, it's my own fault." And George went away.

Noah poured for Clarisse and himself. "Sorry," he said, "I didn't mean to depress you. Besides, breaking up with the Prince is nothing to get depressed about. I haven't slept with him since Valentine's Day two years ago."

"It's something else."

"What?"

Clarisse took her cigarettes from her pocket, tapped one out and leaned over to light it in the candle flame. She drew in deeply and let the smoke slowly escape her mouth. She tossed her head ever so slightly and looked directly into Noah's eyes.

"The reason I'm depressed," she said, "is that I've been putting off talking to you, and I can't put it off any longer."

Noah smiled. "Family business? All family business is nasty."

"Not family business."

"You want to borrow money?"

She laughed. "No, I have my food stamps and a quartz heater."

"What then?"

She tapped the ash off her cigarette and took a deep breath. "I need to ask you a few impertinent questions."

Noah sat casually back in his chair and took a sip of the wine. Clarisse leaned forward, resting her arms on the table.

"I need to know everything about Jeff King."

Noah laughed. "You mean, my relationship with Jeff King."

"Yes."

"Well, how much do you know already? You've obviously found out that I knew him."

"I knew he lived with you on Queensbury Street eight years ago." Noah looked at her with surprise. "Nobody told me that directly," Clarisse explained. "I just put together a number of stories I heard—from Angel, from the mysterious Margaret, and so forth. I know he stole your jewelry and your Rosenthal china, and I know you forgave him." He grinned now. "And I know he came to see you on the afternoon before the Garden of Evil party."

"You know a lot," he said, without any animosity or hesitancy, "now what else do you want to know?"

"What did he want when he came to see you that Saturday afternoon?"

"He wanted to stay with me. I said no, I didn't have a spare room, so then he said he'd sleep with me. I respectfully declined. Jeff had a kind of paranoia—he wasn't comfortable unless he knew he had a place to stay. It's a kind of insecurity that goes with having been a foster child, I guess—always moving from one place to another. He couldn't even think straight unless he knew where he was going to end up for the night. It's a pity he never went to jail—I think that's the only place he would have been really at ease. Anyway, when it got through to him that I wasn't going to let him stay at the house, he tried to sell me drugs. I told him I didn't use drugs. That's not quite true, but I would never buy anything off him—I didn't trust him, and I didn't trust his drugs. He asked if the Prince was home and could he see the Prince. I said the Prince wasn't there and I didn't know where he was or when he'd be back. Then he went away and that was that."

Clarisse said nothing for a moment, but appeared in deep troubled thought.

"What's wrong?" asked Noah. "Don't believe me?"

"I do believe you," replied Clarisse, taking a sip of wine. "It's just hard to believe that you and Jeff King were lovers."

"Lovers implies equality," said Noah. "I *kept* Jeff King. He was nineteen and I was thirty-one. He ran around—I ran around too, of course, but when *he* ran around he charged—and he *stole* from me. And what was worse, he stole from other people in the apartment building. After Jeff I learned my lesson, and never got involved with anyone twenty years younger than me again. Jeff didn't care about me, he cared about money, and drugs, and a roof. He drove me to tears and Valium. When he left, I didn't care—I didn't owe him anything. He started coming to Provincetown about five years ago, ostensibly to get a tan, but I suspect really to deal drugs. He was dealing even when he was living with me, but it was just small-time when you compare it to what he'd been doing lately. I didn't mind small-time stuff, but big-time stuff—you have to get involved with all sorts of nasty people. I'm not interested in having big-time dealers for friends—they have short life expectancies. Just when you get to know one of them, he dies in some very messy way. The Prince changes dealers every year—they get used up fast."

"They all die?"

"Or disappear. Most of them just disappear. And there's nobody to tell you where they went."

Their appetizers were brought; they paused in the conversation to eat.

Noah finished his first, and asked, "Anything else?"

"Yes," replied Clarisse after a moment. "When did you find out Jeff was dead?"

"About an hour after you found him, I guess. The Prince told me."

"The Prince! How did he find out?"

"He heard it on the meat rack."

"But you two left the party early. I assumed you went home to bed."

"We came home, and *I* went to bed. But the Prince was speeding his spikes off and couldn't even sit down, much less sleep."

"And as soon as you found out Jeff was dead, you took off for Boston."

Noah laughed. "It wasn't cause and effect. That trip had been planned. I was meeting Calvin Lark for breakfast at the Swiss Miss in Brookline and we were going over business that I had been putting off for a long time. I left very early to miss the heat and the traffic."

Clarisse sighed and shook her head. "You don't know how relieved I am. . . ."

"This has been bothering you? You thought I had something to do with Jeff's death?" Clarisse shrugged. "I'm not the killer type. In a way," he said, smiling, "it's a fault. You know, I just stop caring. I'm not a man of strong passions. I don't even get bitter. After he left me I never even thought of Jeff unless he was standing right in front of me. Jeff King was a stranger who knew my name, that's all. He's dead and I wish I could say I was sorry, but the fact is I'm not. I just don't care."

"And you forgave him for what he did to you?"

"Oh," shrugged Noah, "I got the Rosenthal back, you know. . . ."

Chapter 30

Valentine took a long swallow of his beer and looked about the bar. It was only nine-thirty but Back Street was already crowded; the rainy streets had driven the men inside. He rested an elbow on the bar and hooked the heel of his boot over the brass footrail. Rock music pounded from the half dozen speakers suspended in the corners of the large basement bar. He looked to his right and left and then straight ahead, trying unobtrusively to check his image in the framed mirrors attached to the bare brick walls. The red track lights, he decided, were sufficiently low to mask his burn, and if he were careful which profile he presented, he might even give the appearance of having a healthy tan.

He finished his beer and signaled the bartender for another. It was brought immediately and payment refused with a friendly wave. Valentine left a dollar on the bar and turned back to the crowd. He checked a wall clock. He'd been in the place an hour, alternately wondering why no one was cruising yet and what on earth he was going to say to Terry O'Sullivan when he saw him again. Though he gave equal time to each consideration, the latter was patently the more important. But for the life of him he couldn't think now what more he wanted to ask Terry. He was certain that the death of Ann Richardson had been an accident—an accident for which Terry O'Sullivan was morally responsible, but an accident nonetheless. If he needed no other information from Terry

O'Sullivan, he could at least pound the fact of that responsibility into the man's selfish skull.

Valentine was no longer certain that Clarisse was on the wrong track when she insisted on searching for the pusher's killer. When Valentine discovered that Terry O'Sullivan had been withholding information regarding Jeff King, he realized that others might be lying too—and not only was the killer obviously still at large, he might be in this very room.

Valentine pushed away from the bar and filtered through the crowd of men. He recognized many from having served them at the Throne and Scepter. To some he nodded friendlily, and received only puzzled brief acknowledgments in return. He sighed. His mystique as a bartender evaporated once he walked out from behind the bar.

He swung around a pole to linger in one of the back areas where the crowd was less dense. His eyes shifted to the entrance across the room as more and more men came into the bar, each of them looking about with feigned disinterest as he paid the cover.

None of these men drew Valentine's mind entirely from thoughts of Terry O'Sullivan, until he saw, under the red spot by the front door, a shiny leather motorcycle cap. When the brim was raised he saw beneath it a drawn dark face with a neatly cut beard and an overfull mustache. The man was short and slender. He wore a ruby-colored T-shirt beneath a black leather vest with chains sewn onto the left shoulder, studded black wristbands, black leather chaps over worn jeans, and heavy black boots with spurs.

A group of animated chatterers suddenly moved between the man and Valentine. Val moved quickly to one side and regained his view; they locked eyes, but the other man's expression did not change. He held Valentine's gaze long enough to show he meant business, then winked slowly and turned away.

Valentine tightened his grip on the can of beer, crushing in the sides. Beer sloshed out and down one leg of his jeans. He swore, dropped the can into a trash barrel and yanked his red bandana from his back pocket to daub at the wet denim.

"Forget that shrimp in leather, lust has been dropped on your doormat."

Valentine turned to see Clarisse standing beside him, shaking the excess water from her umbrella.

"Where did you come from?" Valentine asked and shoved the bandana back into his pocket.

"I was two people behind Suzy Sawed-off, but you only had eyes for him. I've been to six different bars looking for you. Are you sure that's real leather? It looks suspiciously like Naugahyde to me."

"I can *smell* real leather, even from that distance. And what's this about lust on the doormat?"

He raised two fingers to a passing waiter. The man went off to get the order. Clarisse rested the umbrella against the wall, leaned against a pole and lighted a cigarette. She told him that Axel Braun—alone—had become the new tenant for the rental apartment.

"Scott, apparently, is out of the frame. You have a clear path to Axel's arms."

"I wouldn't mind being faithful for a week or two," mused Valentine. "I've been watching you in the throes of domesticity with Officer Montalvo, and it's set me to thinking. . . ."

"Don't think too hard. I had to have a little talk with Matteo tonight."

"What about?"

"Jealousy. He got jealous of Axel when he saw him practically naked in the kitchen within arm's length of me. I don't like jealousy—in fact, I don't *allow* jealousy. It's sweet," she smiled, "but it really turns me off." She frowned.

"How did Matteo take the lecture?"

"Like a man."

The waiter brought their beers. Clarisse looked at the label, grimaced, handed it to Valentine, and sent the waiter off for a scotch on the rocks.

"How did it go with Noah?" Valentine asked. "Any bombshells?"

"Wet fuses all the way. Everything was plausible. But I can't say I'm sorry to find out Noah had nothing to do with Jeff's death." She briefly recounted the conversation with her uncle.

Valentine said nothing for a moment, and appeared only to be checking out each new arrival. Then he asked, "Do you

160

believe that Noah had been planning that trip to Boston? Even though you know that he had changed his will and insurance policies to cut out Jeff King, who was already dead?"

"Of course I believe it," said Clarisse lightly.

"No you don't."

"I don't know what to believe," she admitted.

"Maybe it is just coincidence, but don't you think it's murky that two people who came in direct contact with Jeff King that day pulled disappearing acts with the first light of dawn?"

"Noah," said Clarisse with a frown. "But who was the other?"

"Scott DeVoto. Axel told me that he drove off that morning also." Valentine was looking past Clarisse as he spoke. He took a swallow of one of the beers he held. "Scott seems fond of exits and entrances."

"What do you mean?"

"You just told me that Axel was alone on Kiley Court."

"Scott's in New Hampshire, at one of those lakes with a summer-camp name."

Valentine smiled and tilted his can to point across the room. "Then the nearly naked Mr. Braun is in for a surprise."

Clarisse spun around and saw Scott, his tank top removed and draped over his shoulder. His hair and face were damp, and he was nodding in time with the music as he swayed on the edge of the dance floor.

She looked at Valentine. "Axel specifically said Scott was in New Hampshire."

Valentine sighed. "I was contemplating hot nights cuddled up to Axel braiding his chest hair with my teeth. Life is one cruel disappointment after another."

Clarisse sighed too. "I hope those two are *not* going to stage impromptu productions of *Who's Afraid of Virginia Woolf* in the courtyard."

"The Prince may upstage them. If what you say is right and Noah is going to kick him out, I'm sure the Prince can beat anybody at the ungracious-jilted-woman game." He raised the second beer can in another direction, and Clarisse looked.

The White Prince was just coming out of the ladies' room. He paused to cinch in the belt of his trench coat before he headed for the door.

"Did you see him in here before?" asked Clarisse.

Valentine shook his head. "I—" he began, but Clarisse jabbed him in the ribs and cocked a thumb over her shoulder.

He peered cautiously around her, and saw Terry O'Sullivan staggering up to the bar and gripping the edge of it for balance. He ordered something, and Valentine was very surprised to see that the bartender poured out of a liquor bottle.

"Everybody in the world you didn't want to see tonight is in this bar!" exclaimed Clarisse. "I thought Terry didn't drink."

"That's my doing," said Valentine. "I started him off this afternoon."

"Poor baby must be unhappy. Did you send him packing again?"

"I've saved the best for last," replied Valentine, and he told her what Terry O'Sullivan had revealed to him at the Throne and Scepter.

"No sympathy! No sympathy!" cried Clarisse, when she heard the end of it. "I hold that man personally responsible for Ann Richardson's death!"

"Shut up!" said Valentine, peering over her shoulder. "He sees us and he knows we're talking about him."

Clarisse whirled around and glared at Terry O'Sullivan. "I don't care what he thinks," she said, turning back around. "Oh God," she said, in another tone of voice, and nodding in yet another direction. "Here comes your friend the midget. With all those chains, he looks like Marley's Ghost come to cruise."

"He can pull aside my bed curtains any day."

"Go talk to him."

"I can't. Terry O'Sullivan's on his way over."

Clarisse smiled maliciously. "Let *me* talk to Mr. O'Sullivan for a few minutes. By the time I get through with him, he's not going to be in any shape to make dates."

She turned smoothly around to intercept Terry O'Sullivan. He was almost directly in front of her, but he stumbled awkwardly. His glass dropped from his hand and smashed on the floor. His cheeks and forehead were flushed a ghastly red-purple, his eyes were wide with fright, and his mouth gaped as he gasped for breath.

Clarisse stepped aside, and jerked Valentine up beside her.

Terry O'Sullivan struck his breast once with his right fist, and then crashed to his knees at Valentine's feet. He tumbled backward and hit his head on the cement floor. The sickening crack could be heard above the disco.

Valentine clutched Clarisse's forearm and yanked her down onto the floor next to Terry.

"Give us room!" he shouted to the men who had quickly crowded around. They shuffled back a little.

"What are you doing!" cried Clarisse as Valentine ripped open Terry's shirt, scattering buttons.

"You took the CPR course, not me!" he snapped as he pulled Terry's arms flush with his sides.

Clarisse quickly angled Terry's head back into an arching position, raised her clenched fist and brought it down powerfully against the man's unmoving sternum.

PART IV

Prostitution through the Ages

Chapter 31

Four days later Valentine and Clarisse were somewhat recovered from the trauma of seeing Terry O'Sullivan die on the floor of the Back Street bar. The paramedics who arrived eleven minutes after Terry's collapse concurred that Clarisse had done, and done correctly, all it was possible to do—and Terry had still died.

Clarisse went to work the next day, however, and the next and the next, but when Valentine's own day off came around, she begged Beatrice to let her off, without pay if necessary. Beatrice agreed, and early in the morning Valentine and Clarisse walked out to Race Point Beach. They had lain several hours on adjoining towels, reading and napping, and not done much talking. They had swum together, strolled along the beach together, and together they had avoided talking about Terry O'Sullivan.

It was now almost noon. The day was hot and the sun high in the clear sky. The ocean lapped gently not twenty feet from them, and they lay just at the unofficial demarcation line between the gay beach and the straight. Clarisse noted that the division was not hard to see. To her left the men and women were mostly either pale or red, unused to the sun; the men had fleshy unkempt bodies and the women sprawled and shrieked in a desperately unattractive manner. To her right, however, the towels and blankets were adorned with browned, toned, or muscular bodies carefully oiled. Men and women read, slept, or conversed in whispered voices that never rose louder than the waves. If there was laughter it was deliberately musical. Even the music was decisive: to her left was old disco and to her right new wave. Beyond the sunbathers the dunes rose starkly against the washed-out sky. Gay men alone or in pairs now and then went up the sandy mounds and disappeared over the crests. Heads bobbed up occasionally but then dropped quickly out of sight again. If earnest cruising were not sanctioned in Provincetown proper during the day, it was a constant activity in the dunes of Race Point.

Valentine sat up and peeled off his tank top. He pulled it through one belt loop of his cutoff jeans. His sunburn of the previous week, by the careful nurturing of his less affected side, had evened out and deepened. He wiped a sheen of perspiration from his face and lay back on his elbows.

Clarisse held her hand up before her face, flexing the fingers before the large lenses of her gray-tinted sunglasses. "It's still bruised," she said. "I'm surprised his chest didn't cave in."

"You were wonderful," sighed Valentine. "Clara Barton must be smiling in her grave."

Clarisse grimaced. "I'll bet Terry's not smiling in his, though. I'm just glad I didn't faint at the sight, like that friend of yours in the leather and chains."

"Listen," said Valentine, "I want to apologize to you."

"About what?"

"About making light of the trauma you suffered finding Jeff King and Ann Richardson dead at your feet."

"Now you know what it feels like."

"It doesn't feel good," said Valentine. "It's not nice to see somebody you know die. I keep trying to tell myself that Terry was responsible for Ann Richardson's death—and so he deserved what he got."

"Do you think he died of guilt?"

"No. He died of a massive heart attack."

"*That's* what I don't understand," said Clarisse. "He was too young to die of a heart attack."

"For all his yammering about the care and feeding of the gay body and soul, I don't think Terry took very good care of himself. Did you ever have dinner with him? All hamburgers and French fries. *Always*. And lots of milk shakes—he had an ulcer. He never went to the gym, and when he was in Boston he put in a twelve-hour day. All work and no play makes Jack a cardiovascular statistic."

They were silent for several minutes. The subject was not yet one they could treat lightly.

"I'll accept your apology," said Clarisse. "On one condition."

"What?"

"You'll listen to me for five minutes without getting upset, without interrupting, and without rolling your eyes."

Valentine rolled his eyes. "All right."

"Jeff King," Clarisse said, "went swimming in the bay, and despite what the autopsy said, hit his head on some pilings, lost consciousness, and drowned. Ann Richardson mixed MDA and angel dust, jumped in the pool, and didn't come up the third time. And Terry O'Sullivan, seeing your gorgeousness once more from across a crowded room, gave up the ghost and tumbled dead at your feet. Right?"

Valentine nodded once.

"I still can't get over the feeling they're connected," said Clarisse.

"Why?" asked Valentine politely.

"Well, if for no other reason than that they all knew one another. When was the last time you heard of three people who knew one another dying within three weeks of one another—and the deaths weren't connected?"

"Ann didn't know Jeff," protested Valentine, "she said she didn't."

"All right," admitted Clarisse, "but there was a connection there, through Terry. He knew both of them. Now here's what I'm getting at. Just suppose that all three of these deaths were murders—I'm not saying they were, I'm just saying suppose they were."

"All right," said Valentine, "I'll close my eyes and make believe."

"Thank you," said Clarisse. "Now, you and I knew all three victims—or at least I knew all three victims. Not well, but I knew them."

"Got it," said Valentine. "Now what?"

"Doesn't it also make sense that I know the murderer too?"

169

"I'll accept that," replied Valentine. "But check your assumptions, Lovelace."

"What assumptions?"

"You're assuming that only one person killed all three. What if two people got together and killed all three? Or what if one person killed Jeff King and somebody else killed Ann and Terry? Or what if five people killed Jeff King—I think we could find five who would have wanted him dead—and two killed Ann and Terry? Or—"

"You promised to listen!"

"I am listening, and I'm not making fun of your idea. I have to make another apology. I've been coming down on you for all this talk about murder and the 'person who did it'—but now I think you may be right. I'll certainly admit that Jeff King was murdered, and that it looks very strange that Ann Richardson and Terry O'Sullivan should die within the next three weeks. I'm not sure they were *all* murders, or if they were, that one person did it—but I think that, on the whole, you were right and I was wrong." He lowered his head on his chest with a jerk, as if the confession had cost him.

"So what do we do now?"

"I don't know," replied Valentine.

"It's still June," said Clarisse, "and I've seen three dead bodies. Is this going to continue?"

"You'd think," said Valentine, "that we could have figured this out by now. Maybe the sun is bleaching our brains," he said, glancing up into the sky and squinting.

They were silent for several moments.

"Is Matteo on duty today?" Valentine asked. "Desk duty, I mean."

"Yes," said Clarisse. "Why?"

"Do you think he would let you look at the autopsy reports again?"

"No, he'll kick and scream and rave."

"Bribe him."

Clarisse lifted her glasses and peered at Valentine. "Hey," she said, "this means you're on my side now, doesn't it?"

The police station was crowded, and Matteo protested in whispers. But he brought out the files—including the one pertaining to Terry O'Sullivan—and slipped them into an issue of *Life,* which Clarisse took into the ladies' room. She brought them out twenty minutes later, having sat in a stall, read each of them through twice, and taken a few notes.

Meanwhile Valentine went by the Plymouth House where Angel Smith was in frantic preparation for Friday evening's pageant. He stood at the back of the Amaretto Room, and she rushed toward him. He braced himself against the doorframe, but she stopped in time.

"You're coming, aren't you!" she cried.

"I want to reserve a table for two," he said, smiling.

"Sure," Angel replied, looking about the room. "Which one?"

Valentine pointed to an enormous round table in the back corner that would seat at least eight, and replied, "That one."

"How many comps is this going to cost me?" she grimaced. "You know how much it takes to put on a show like this? The uppers alone . . ."

Valentine took out his wallet. "I'm buying eight tickets. Tonight we're paying guests. Just put a reserved sign on the table."

Angel nodded and sighed. "You would have made a great Du Barry. . . ."

Chapter 32

Valentine and Clarisse, with Noah between them, arrived at the entrance of the Amaretto Room just after seven o'clock that Friday evening. Clarisse remarked that she had never before been an hour early for anything in her entire life, and the two men believed her. The evening was balmy—as perfect an evening in fact as Provincetown weather is capable of: warm, but without dampness, with a slightly pungent breeze. The moon was waxing and the night sky was cloudless. A large banner had been suspended over the door of the cabaret announcing:

PROSTITUTION THROUGH THE AGES
SPECTACULAR TABLEAUX D'ART
2 NITES ONLY

Below this, on either side of the garland-festooned doorway, were posted four torchbearers. They wore loincloths—all four costumes stitched together wouldn't have mopped up a glass of spilled milk. Their bodies had been covered toe to head in gold paint and their hair had been dyed a color that matched the glass of their electric torches. They were admirably stony-faced, and refused even to acknowledge their friends in the crowd.

"Angel must have slipped them some incredible downs,"

whispered Clarisse, but Valentine did not hear her. He was too busy mumbling the torchbearers' names and bedroom predilections to Noah.

The door would not open until seven-fifteen, but already a sizable crowd had assembled, many with their pink invitations in hand. Everyone, it appeared, had donned his summer finery, the evening being the equivalent of the opening of the opera in Boston. White cottons predominated among the men and pastel silks among the women. Valentine and Noah wore white linen suits with open-collared shirts beneath. Clarisse wore a knee-length white forties-style dress beneath a silver bugle-beaded waist-length jacket. She straightened the padded shoulders and touched her hand to her hair, which was up in a smooth victory roll. She looked over the crowd expectantly.

"Looking for someone special?"

"I left an invitation in Axel's mail slot. I was hoping he might show up early."

"Why didn't you just leave it under Daniel's pillow?—then he'd have been sure to get it."

The door of the Amaretto Room swung suddenly open and two women in crimson togas emerged carrying long-necked trumpets. They raised them high and blasted a shaky fanfare right in Clarisse's ear. They bowed and disappeared inside again. Valentine, Clarisse, and Noah were the first to enter, and immediately seated themselves at the large round table.

Large paintings of full-bodied women languishing on beds, couches, chaises, grassy banks—and even on kitchen tables— had been hung on the walls. Small cut-glass chandeliers had replaced the usual fixtures. Enormous baskets of red and yellow flowers had been placed at either side of the stage, a raised platform fifteen feet wide at one end of the room. The red velvet curtain had been replaced with a deep purple, gold-fringed one.

The three of them looked about, nodding to acquaintances. "It really is amazing," said Clarisse.

"What?" asked Noah.

"I work all day long in that shop that is boycotted by anyone with taste. My nights I spend at home in bed reading—"

"Or just in bed," interjected Valentine.

"And the only people I see are you and Val and Matteo, and yet I think I know half the people in this room. Where did I meet them all?"

"You didn't meet them," said Noah. "It's just that you see them on the street every day. I suppose that passing somebody twenty times in twenty-four hours on the same sidewalk constitutes a kind of introduction."

A small orchestra had set up to one side of the stage. The seven musicians tuned their instruments and then struck up a medley of obscure show tunes—but never so obscure that some man in the room didn't know *all* the lyrics and insist on singing along in a cracked voice.

"I thought Matteo was on the door this evening?" said Valentine.

"He doesn't come on duty until . . ." Clarisse faltered. Valentine and Noah were staring past her with expressions of surprise and alarm. ". . . eight," she finished, and turned slowly, following their line of vision. "Oh my," she whispered.

Margaret stood a dozen feet away. Her hennaed hair glistened in the red spotlight over the door. She wore a black strapless gown that swept the floor. Her eyes wandered over the crowd, and when she saw Clarisse she smiled and headed over to the table.

"Are you saving these places for anyone special," she asked, "or may I join you?"

Silently, stunned, they motioned for her to sit.

She greeted them each in turn, with an increasingly puzzled glance. With a little smile and a slightly furrowed brow she asked, "What's new?"

Valentine and Noah exchanged uneasy glances.

"What a surprise to see you here, Margaret," said Clarisse at last. "We didn't expect to see you back in town this summer."

"Oh, yes, I'm sorry I didn't get a chance to say good-bye, but I knew I'd be back this weekend and I'd probably run into you again. And so I have."

"You knew you were coming back?" asked Valentine.

"Oh, yes, Ann and I bought tickets for the show as soon as

we heard about it. In fact, we didn't know where we'd be staying, so we just planned to meet here. More romantic, I guess. Do you think there'll be room for her at this table too?" She looked round at the four remaining chairs.

No one said anything for several moments. Then Clarisse smiled weakly, laid her hand on Margaret's forearm, and said, "Why don't you come backstage with me for a few minutes and say hello to Angel before the show starts? I know she'd want to speak to you."

"Oh, I'd love to!" Margaret said, rising.

Valentine and Noah said nothing.

Clarisse led Margaret up to the stage, pulled back the curtain at one end of it, and stepped behind. They made their way through the wings and into a long hallway. In the small dressing rooms on either side of the corridor pandemonium reigned. Men, some in near-full costume and others hardly dressed at all, flung themselves about, screaming for wigs, pancake number five, bras, safety pins, Quaaludes, glue, and glitter. A young man stood peering intently into a mirror and industriously painting his nipples blue. He looked up and smiled. Clarisse leaned into, but did not enter, each room searching for Angel. A thin emotionally frazzled little man of middle-age appeared at the end of the hall and waved a clipboard ineffectually about. "Twenty minutes to curtain!" he shouted. First came a prolonged stereophonic screech, and then a chorus of obscenities was showered upon him from up and down the hallway, many ending with the words, "you wimp!" The man rushed toward the stage. Clarisse caught a glimpse of the Prince racing from one room to the other wearing a blond wig that was only half combed out, a corset, panty hose, and lethally high spiked heels. He pushed past, according Clarisse a brief "Hello" and Margaret a sharp look of surprise. He left a wake of lavender perfume.

Clarisse and Margaret found Angel in the last room. She sat on a little wicker bench, the legs of which had been reinforced with steel straps, at one end of a fifteen-foot-long makeup mirror. Also in front of the mirror sat a line of seven men in towels, jockstraps, women's foundation garments, and G-strings. The air was misty with powder and pungent with hairspray. Angel was frantically searching through a drawer before her. Her blond hair was done up into a cloud of lacquered waves. A full-length dressing gown concealed her costume beneath.

She slammed the drawer shut and wailed, "I've lost my blush! I can't go on! Call off the show!"

There was a flurry of movement among the men, and in another moment half a dozen small containers of rouge flew through the air and clattered on the table before Angel. Without comment she opened them all, chose one and applied the color generously to her cheeks.

Clarisse and Margaret moved about until they stood behind Angel, who flicked her eyes up to see them in the mirror. Her mouth dropped open when she saw Margaret. She turned to face them and looked to Clarisse for an explanation.

"*Why* is everyone looking at me like that tonight?" demanded Margaret.

Angel mustered a smile that faded quickly. Then she looked at Clarisse. "She doesn't know, does she?"

Clarisse shook her head.

Angel stood up, and motioned for Margaret to take her place. Margaret, with some apprehension, did so. "Do I want to hear this? Is my dress in bad taste?"

"Margaret," said Clarisse evenly, "Ann is dead."

A weak smile pulled at Margaret's mouth and her eyes narrowed with confusion. All the chatter from the men at the makeup table suddenly stopped, but no one turned to stare. They leaned forward into the long mirror, and the painting went steadily on.

"When?" said Margaret, confused, not knowing what to ask first.

"The night you left," said Clarisse, and told about the drowning.

"But Ann was a great swimmer!"

"At first," said Angel, "the police thought it was an accident, but after the autopsy, they changed it to suicide."

A man in a blue-sequined jockstrap reached around Clarisse and handed Margaret a glass of water.

"Good swimmers don't commit suicide by drowning," protested Margaret. "Good swimmers swim automatically, there's no way for them to drown. The only people who commit suicide by drowning are people who don't know how to swim."

"It was the drugs," said Clarisse, and explained further. Margaret had nothing to say.

"Why did you leave town so quickly?" asked Clarisse.

"I had planned to leave that afternoon," said Margaret, obviously thinking of something else. "But Ann kept pleading with me to stay longer, so I didn't get off until one. She was depressed—very depressed. So I called up her boss—you know who I mean?—and asked him to come over and take care of her for a little while. They had had a fight that afternoon, or the day before—I can't remember—but he was the only person I could think of to call."

"Terry O'Sullivan came to the house that night?" said Clarisse, surprised.

"I don't know," said Margaret, still distantly. "He said he was coming right over, but he wasn't there by the time I left. I *had* to go, I was getting a ride with a friend. We drove all night."

She sat for a few moments despondently, then looked up suddenly at Angel. "Oh, you have to get ready to go on," she said weakly. "And I'm in your way."

"No, you're not," replied Angel. "Sit there as long as you like."

"No," she said, "I want to go back to the table." She stood, only a little unsteadily.

"Will you be all right?" asked Clarisse. "Do you want me to call a taxi?"

"I'm staying about two houses down. No, I don't want to be alone now," Margaret said. "I think I want to see the show. I'll be all right. I'll save my uncontrolled weeping for later."

Chapter 33

"How'd it go?" Valentine asked in a whisper. "She seems to have taken it pretty well. Are you sure she didn't know about it already?"

"Hush!" cried Clarisse. "Of course she didn't know about it already. I know a broken heart when I see one."

Valentine and Clarisse sat with their heads together, talking in low voices. Margaret sat on the other side of Clarisse, silently smoking and staring sightlessly at the stage. Axel Braun had come in some minutes before and was chatting with Noah.

"I think she's in shock," said Clarisse, "but she insisted on staying."

"Ask her if she wants to come back with us tonight," suggested Valentine. "She probably shouldn't be alone."

"She also probably shouldn't spend the night at the place where she last saw Ann alive."

"Hmmmm," said Valentine, nodding. "Did she say anything else?"

"That her last name is Richardson, same as Ann's."

"Angel was right then," said Valentine. "I thought she was mixed up." Clarisse also told him what Margaret had said about Terry O'Sullivan's possible visit to the house on the fatal night. "So what do you think?" said Valentine. "Terry showed up after Margaret was gone, got Ann stoned, which wouldn't have been hard in her frame of mind, and then—"

Clarisse made a downward movement with her hand.

"Plunked her under and left. In which case, mixing the dust and the MDA would have been deliberate."

"Well," said a loud voice just behind them, "here I am!"

Everyone looked up. There stood Scott DeVoto, smiling blandly. He flung a pink invitation onto the table. "Thanks for the invite," he said to Axel sarcastically.

Valentine flipped the invitation over. The address on the envelope was in Clarisse's handwriting. Clarisse smiled innocently and plucked a piece of imaginary lint from her padded shoulder.

Axel didn't get up. "I didn't send you an invitation," he said.

"I know," said Scott, taking a chair next to his lover. He looked coldly at Valentine and said, "I'm not breaking up a heavy date, am I?"

"I'm escorting Clarisse," said Valentine impassively.

"Don't start," said Axel.

"Order me a gin and tonic," said Scott. He looked at Clarisse, and said, after a moment, "You were great in Back Street when you did the CPR on that guy."

"Thank you."

"Too bad it didn't work. He would have owed you a zillion favors for that one. It was a heart attack, wasn't it?"

Clarisse nodded.

"I don't think I could do CPR on anybody unless he was cute," said Scott. "What if he had had hepatitis or something —or he could have had trench mouth. If he had a bad heart, he shouldn't have been doing so much coke."

Scott's drink arrived, and Noah ordered another round for everyone else.

"Coke?" asked Valentine.

Scott nodded. "That night in Back Street the dead guy was doing coke."

"How do you know?" asked Axel.

Scott shrugged. "I was in the ladies' room drying my hair, and in one of the stalls there were these two guys snorting their brains off—at least one of them was."

"How did you know it was Terry O'Sullivan if he was in one of the stalls?" asked Clarisse.

"Because he was using a dollar bill to snort it—cheapskate! —and he dropped it and it rolled out from under the stall and

179

he had to come out to get it. I couldn't see the guy he was with, except for the top of his head. Maybe this Terry guy was trying to kill himself. Bad heart, and a few good lines of coke and . . ." He snapped his fingers for emphasis.

"He wasn't the type," said Valentine.

"I don't know," said Scott. "He was kind of a trouble-maker. Real Captain Wimp. He was always turning up."

"Like where?" asked Clarisse.

"Here and there," said Scott casually. "Like at your house the night that woman drowned in your pool."

Valentine was astonished. Clarisse glanced anxiously at Margaret, who was looking in another direction.

"Just pretend I'm not here," said Margaret, calmly tapping the ash from her cigarette.

Clarisse addressed Scott. "So that was you lurking in the courtyard that night."

Scott took a long swallow of his drink and glanced mean-ingfully at Axel. "I wasn't *lurking*," he said. "I was spying."

"On Terry O'Sullivan?" asked Noah, puzzled.

"No," said Scott, sniffing. "On Axel, of course."

"But Axel wasn't staying at the house then," persisted Noah.

Scott looked at Noah as if to say, *Who are you?* Then he answered, as if the entire table had asked the question. "Axel and I had a fight that night, out in front of the Throne and Scepter. Nothing serious—just the usual." He turned to Axel with a smile. "You thought you left me at the A-House, but you didn't. I followed you to Back Street." He smiled. "I saw you meet him"—he jabbed a finger toward Valentine—"and I wondered if you'd actually go home with him, so I went over to Kiley Court and waited."

"How did you know where I live?" demanded Valentine.

Scott smiled even more sweetly. "I had followed Axel there once before."

Axel looked at Scott darkly.

"So it was very foggy and I hid on the other side of some bushes by the restaurant across from your house. There's even a bench there, so I sat down and smoked a joint and waited."

"Why the hell did you do all that?" demanded Axel.

"I felt like it."

"And you saw Terry O'Sullivan?" asked Valentine.

"I saw the whole world and his dog," said Scott. "First I saw you," pointing to Noah, "and you were with somebody, but I didn't see who it was."

"That was the White Prince," said Noah.

"Then I saw you," nodding toward Clarisse, "and I heard you talking to somebody in the courtyard—women's voices. Then another woman came out with a suitcase. Then Axel and *that one* came back." He looked sourly at Valentine.

"I take it you had a notebook and stopwatch," said Clarisse.

"No," said Scott. "Actually, I'm not sure if I got the sequence right or not. I was stoned. And the only thing I really cared about was whether Axel was going to show up or not."

"But where was Terry O'Sullivan in all this traffic?" asked Valentine.

"What does it matter?" replied Scott rudely.

"It matters," said Clarisse emphatically.

"He got there after you did, I guess," said Scott to Clarisse. "I remembered him—I had seen him before. I thought he was coming to do a threesome with Axel and your *escort.*"

"He was not," said Valentine forcefully.

Scott looked away. "Anyway, he arrived right after the woman with a suitcase came out."

"That was me," said Margaret, turning toward Scott. Clarisse hadn't been certain that she was listening to any of this.

"Maybe," said Scott. "Women all look alike to me."

"How long did Terry stay?" asked Valentine.

Scott ignored the question.

"How long did he stay?" repeated Clarisse.

"A while," said Scott. "I don't know."

"He had to have left before we got back," said Valentine. "I sure didn't see him."

"No," said Scott, "he was still there when you and Axel came in."

"You're sure of that?"

"Yes," said Scott, "I'm sure, because I saw him go in, then I saw you go in, then I waited for a while. He eventually came out, and then I got tired of waiting around so I left."

"Do you know where he went?" asked Clarisse.

"Sure. The Boatslip—Room 231."

Everyone looked at Scott in surprise.

"You followed Terry too?" asked Clarisse.

Scott smiled. "He seduced me."

Valentine gave a short laugh. "Terry O'Sullivan couldn't have seduced the Whore of Babylon."

"He didn't have very nice things to say about *you*," he said to Valentine.

"I need a 'lude," said Axel to the company in general. "Anybody got a 'lude?"

Clarisse put her hand on Valentine's arm. "Did he say anything about Ann?" she asked Scott.

"Ann who?"

"The woman he probably killed," said Noah.

"Why would he have killed Ann?" asked Axel.

The houselights had dimmed. The medley of show tunes had finished, and now the music came up loud and stirring.

"Well," said Margaret loudly over her shoulder, "it might have had something to do with why Terry O'Sullivan killed Jeff King."

Valentine's mouth dropped open, and Clarisse's importunate question was lost beneath the applause that greeted Angel Smith's head, thrust between the curtains.

Chapter 34

An intense white spot was focused on Angel's face, which was all of her that was visible before the curtain. She smiled bravely, and in a melancholy unaccompanied treble sang:

> The ballroom was filled with fashion's throng,
> It shone with a thousand lights,
> And there was a woman who passed along,
> The fairest of all the sights.

She whistled an exquisite trill.

> A girl to her lover then softly sighed,
> There's riches at her command,
> But she married for wealth, not for love, he cried,
> Though she lives in a mansion grand.

The band struck up again, the curtain parted, the spotlight widened and turned a brilliant gold, and Angel appeared in all the splendor of her costume: a black tuxedo, and instead of a white shirt, a vast expanse of canary feathers. With an even braver smile she began the familiar chorus:

> She's only a bird in a gilded cage,
> A beautiful sight to see,
> You may think she's happy and free from care.
> She's not, though she seems to be.

'Tis sad when you think of her wasted life,
For youth cannot mate with age,
And her beauty was sold
For an old man's gold,
She's a bird in a gilded cage.

A second verse had the same sad introduction, she repeated the chorus twice, and violently rattled the gilded bars of her imaginary cage. There was riotous applause, calls of "Encore!" and, unaccompanied, she yodeled the chorus full volume. When she breathlessly finished, she wiped away her glycerine tears, shoved the microphone just under her lips, and said, "Thank you so much, ladies and gentlemen. Here am I, Angel Smith the Swiss Miss, bringing you tonight the best in Provincetown entertainment, a show that I hope you will find as instructive as it is ornamental. Tonight you and I are going to take a little trip across time. We're going to examine the part that fallen women have played in the comedy we call history. We're going to look at happy women and sad women. We're going to see honors and wealth, deprivation and degradation!" Here Angel's speech was interrupted by cheers. She smiled, then raised her hands for silence. "We're going to see scaffolds and thrones! I'm going to show you bullwhips and crucifixes! There's gonna be diamond tiaras and branding irons!" There was more cheering. Angel danced across the stage. She turned and spread wide her arms. "We're gonna look at deification, excommunication, anathema, coronation, inebriation, and enforced jury duty! I'm gonna show you power, love, lust, gluttony, rage, and jealousy! There's gonna be potato chips and dip, and a good time will be had by all!"

Her feathers shook with her excitement, and the audience applauded madly.

Angel Smith bowed, straightened, and smiled sweetly. "But first," she said in a low voice, "I want to thank some of the many fine organizations that have contributed to make this night possible." She named several bars, guesthouses, and shops, and in a number of cases pointed out representatives of those concerns. A spotlight briefly picked out these persons in the audience, which by now had grown to about four hundred.

"And last and not least, I'd like to thank Beatrice Rowell's Provincetown Crafts Boutique, which during the day is presided over by my very dear friend, Clarisse Lovelace. Clarisse is in the audience tonight."

The spotlight clicked on again and wandered around the audience, but could not pick out Clarisse. Valentine giggled and Clarisse sighed. "They can't get it right here either."

There was a dim voice from the man running the lights from a high platform on the far side of the room. "I can't find her," he called.

"She's in the back," said Angel into the microphone. "She's impossible to miss. Clarisse is the one with the tits that go from here to the Bourne bridge—and back."

The smile froze on Clarisse's face. The spotlight came closer.

"Once I invited Clarisse to a party," said Angel, waiting for the spot to find its quarry, "and her tits showed up five minutes before she did. She has to pry 'em apart to look at her feet. Last week she was voted Sweetheart of the International Brassiere Makers Union."

Clarisse balled her fists, and said in a low voice, "There is a woman at the front of this room who is talking into a microphone about my . . . , my physical attributes."

"Hide 'em," suggested Valentine.

The spot finally found Clarisse, and widened just enough to shine upon her breasts as well as her head. Clarisse nodded in acknowledgment of the wild applause, then stood. She raised her hand, and everyone grew silent. Her face was a smiling frozen mask. She said in a loud clear voice, "I'd just like everyone in the room to know that this is the single most embarrassing moment of my entire life."

She sat down, and the spotlight lingered on her until at last, the applause died away.

"And now," said Angel, "on to the tableaux. We begin at the beginning. Not, as you might think, with Eve. Adam asked for and deserved what he got. He drove Eve to it. We begin with Lot, on a hill overlooking the destroyed city of Sodom." She was interrupted by loud cheers of "Yea Sodom!" and "Right on, Gomorrah!"

The curtain opened to reveal Lot, represented by a young man with a strong clean-shaven jaw, a chest with well-developed pectorals, a washboard belly, and long sinewy legs. He was naked except for a length of shot-silk draped across his hips. He lay in a drunken stupor, while his two daughters—represented by two very tall and very slender men in long flowing wigs and diaphanous spangled gowns, crawled toward him on their hands and knees. On a painted backdrop was Sodom in smoking ruins. The tableau was held for fifteen seconds, then the curtain closed.

The curtain opened again, and the three performers bowed to the applauding audience, Angel introduced them by name, and the curtain was closed. Angel disappeared, the lights came up, the waiters circulated in the audience, and the band played for a few minutes, until the next tableau had been prepared. This was the format for the next hour and a half, with Angel introducing such tableaux as Lola Montez inspiring Lizst, Josephine Beauharnais and Madame Tallien dancing before Barras, the Suicide of Antinoüs, Nell Gwynne leaning over the wall, Henri III and his minions on the way to mass, Hephaestion and Roxanna fighting for the favors of Alexander the Great, the conversion of Thaïs, Mary Magdalene calling for her zebras to be harnessed, Christine Keeler and Mandy Rice-Davies testifying at the Old Bailey, Harriet Wilson locking out the Duke of Wellington, Marion Davies presiding at table at San Simeon, Catherine the Great ennobling Prince Orloff, George the I's German mistresses, the elephant and the maypole, Warren Harding's girlfriend hiding in the closet, Du Barry mounting the scaffold, Miss O'Murphy being painted by Boucher, St. Mary of Egypt lifting her skirts to the ferryman, Moll Hackabout pounding hemp, and Phryne revealing herself to the Greek Areopagus.

At the table at the back of the room, Clarisse was desperate to talk to Margaret, and find out what she had meant about Terry O'Sullivan's being the killer of Jeff King, but the noise, the splendor of the tableaux, and the constant bringing up and turning down of the lights made any protracted conversation impossible. Valentine and Clarisse deep in thought, waited impatiently for the program to come to an end.

At last, Angel appeared and said, "We have one last treat for you. This one we've been saving. Now we present for your edification one of the most notorious women in history, a woman—I may say, a *great* woman, whose insatiable lust for power could be fed only through the prostitution of her tender flesh."

The curtain parted, and there stood the White Prince, dressed as Eva Perón, in a voluminous pale blue gown copied exactly from one the dictator's wife had worn to the opera, with a blond wig pulled back into a tight chignon. A scarlet banner emblazoned with the insignia of Perón's presidency was attached at an angle from the strapless bodice to the waist. It was a startlingly faithful recreation of Juan Perón's mistress-wife.

Anger streaked Eva's face and her body was stiff with barely restrained rage as she glared at another woman who stood before her. This second woman wore a revealingly tight gown of dark blue silk with a butterfly bodice. Several pounds of jewelry adorned her neck, arms, and ears; her bright red hair billowed to her waist. Her stance was accusative, one arm raised and finger pointing at Eva.

"Yes!" cried Angel, from the shadows, "the infamous Argentine dictatoress, Perón's greatest weapon. Behind her back she was called 'Evita Piranha—First Lady of Argen*tuna*.' She ate her way to the top. Here we are shown the famous confrontation between Evita and her arch-rival Lyla Cantanya, the notorious Slut of the Andes! Reminding the First Lady, on the *very night* of Juan Perón's election as president, of the old South American adage, 'Once a whore, always a whore'!"

There was thunderous applause. The White Prince was called out twice, and the second time he was joined by the rest of the cast. They received standing ovations. When she had at last succeeded in calming the audience a little, Angel said, "Thank you, we thank you all. Now please stay and have a good time, and tell your friends to come tomorrow night!"

The stage lights went off, and most of the cast went backstage to change. A number of them, however, came out into the audience to join their adulatory friends. The White Prince, still in costume, came to Valentine and Clarisse's table, and took one of the chairs between Margaret and Scott.

"You were Queen of the Ball," said Noah with a smile.

"You are *right,*" said the White Prince, and with a complacent smile received the rest of the table's praise.

Two waiters appeared with a variety of mixed drinks for the Prince, anonymous gifts from awed spectators. The Prince lined them up before him, and knocked the first one back with a single gulp. He went a little slower on the second.

Angel, still in tuxedo and feathered front, pushed through the curtains and made her way to the table. Chairs were brought for her and she sat to one side of the Prince. He offered her one of his drinks and she toasted him with it, saying, "Brava, Evita!"

Axel got up as if to leave, but Clarisse motioned him down again. "Nobody leaves," she said imperiously. "We have a murder to solve. Margaret, what did you mean when you said that Terry O'Sullivan killed Jeff King?"

Chapter 35

"Are you still playing detective?" said the White Prince languidly, irked that his praise had been a subject so soon abandoned. When no one replied, the Prince turned his gaze to Margaret. "We're on hooks, dear. Do continue."

Margaret lighted a cigarette unsteadily. "Terry O'Sullivan and Jeff King—those are their right names, aren't they?—were lovers. I don't know for how long, but they were. Ann found out."

"Lovers!" Valentine exclaimed. "Terry told me he hardly knew Jeff, that Jeff just supplied him with grass."

"No," Axel cut in. "She's right."

All heads turned to Axel, then back to Margaret.

In a low voice, Clarisse said to Valentine, "Looks like you and I are the only ones who aren't going to be able to contribute to this conversation. Was everybody in town lying?"

"At the Garden of Evil party," said Margaret, "Ann and I were standing there talking to Terry when this man dressed as Cain comes up and latches on to Terry. Well, Terry was none too pleased, and didn't even introduce him. But I saw him slip the man a key, which seemed strange because Terry had just told us he was with you." She nodded to Valentine. "He made a big point about it, too. He said you two had something heavy going. So Terry went away, and Cain came up to us later. Of course, we didn't know him from Adam but he started going on about how Terry was his lover but wouldn't recognize him in public and wasn't he a real bastard and all that sort of thing. And then I realized, when I saw him in the light and saw those eyes—"

"Cobalt eyes," said Clarisse, and Margaret nodded.

"—then I realized it was the same man who had come to the house to see Noah."

"Except," said Valentine, "he was looking for *Terry* when he came to the house, because he didn't know that Terry had moved to the Boatslip. Which also means that Terry didn't expect him."

"And when he saw me there," said Noah, "he thought he might as well try to get a place to stay out of me."

"Yes," said Clarisse, "but Margaret, why did Ann lie to me when I asked her about Jeff King? She said there was no connection between him and Terry O'Sullivan."

"Well," said Margaret, "at that time she wasn't sure, because Jeff could have just been lying."

"Ann had worked for Terry for years—wouldn't she know if he had a lover?" said Valentine.

"Remember how uptight he was?" said Clarisse.

"Yes," Margaret nodded. "Ann told me that so far as work went, he was in the closet. He didn't even come out to her until she ran into him at a bar one night after they had worked together in the same room for eighteen months. And if he had had a lover who was a drug dealer, it's just the sort of thing he *would* have kept secret. But then Ann and Terry had a fight. That was on Tuesday. Terry was being a real pest about wanting to talk shop all the time, and Ann was in town to have a good time, and so I sent her up to his room to talk to him. He was very nasty and said awful things to her, and she told him she knew that he had been lovers with the man who got killed on the beach. He said if she told anybody, he'd fire her, so she kept her mouth shut."

"Ohhh!" cried the White Prince. "Intrigue. I *love* intrigue."

"You're ripped," said Angel. "Be quiet."

Valentine turned to Axel. "Why didn't you say something —if you knew Terry and Jeff were lovers?"

"Because I didn't really *see* anything," replied Axel.

"See? What was there to see?"

"At the Garden of Evil." Axel looked at Scott and then down at his drink, rubbing his thumb about the rim of the glass. "There was the fight between Scott and me, which I'm sure everybody at this table has heard about—"

"I haven't!" cried the Prince.

"Yes, you have," said Noah, "be still!"

The Prince adjusted his bodice with a frown.

"Well, the fight was over Jeff. And when it was over, Jeff jumped over the railing onto the beach. Scott ran off and I couldn't find him, so I thought I'd go after Jeff instead."

"Why did you want to find Jeff?" asked Clarisse.

"Who knows? I was drunk—everybody was drunk."

"I don't think there's been sufficient discussion of my performance tonight," remarked the White Prince.

"Shut up, Prince Valium," said Angel.

"Anyway," said Axel, "I caught up with him, on the beach behind one of the guesthouses. He had just come out of the water. So one thing led to another and . . ."

"There was a lot of knee-crashing in the sand," sneered Scott.

Axel ignored him. "But then that man who was Daniel's date popped up out of nowhere, madder than hell. He pushed Jeff around and called him all sorts of names. Jeff was so drugged up he thought it was funny."

"What did Terry want?" asked Valentine.

"He said, 'This is it! No more! We're through!' Then he said, 'Give me back my room key.'"

"What were you doing during all this?" said Valentine.

"Nothing," said Axel. "I didn't say anything. I get into enough arguments on the home front without worrying about somebody else's love life. So I left. I went to the meat rack."

"To look for fresh pork," snapped Scott.

"No," said Axel soberly, "to look for you. I waited for half an hour. It wasn't till I got back to the cottage and saw that the car was gone that I realized you had already left town."

"Well, well, well," said Noah, leaning forward. "So it looks like Terry O'Sullivan killed his shadow-lover in a jealous rage and then when his secretary made the lethal connection, he dispatched her too. Nasty business."

Clarisse sat back, troubled. "He must have mixed the drugs in the wine, and Ann drank it without realizing what she was doing. You'd think she'd have tasted it, though."

"If you knew how much she had had to drink that night," said Margaret, "you'd know she couldn't have tasted a fistful of Mexican peppers if she'd swallowed them."

"But Terry knew exactly what he was doing," said Valentine.

Clarisse nodded hesitantly.

"Oh," cried the Prince excitedly, "I feel just like Ida Lupino in *I Love a Mystery!*" He paused, then added sadly, "Only there's nobody to arrest. Old Terry O'Sullivan committed the perfect crime, then broke his heart in two in a basement bar. There's no trial to attend. What a pity." The Prince sniffed, straightened the banner across his breast, and took a long sip of his fourth drink. "I can just see how it would have been," he said, looking dreamily at the ceiling: "I'm called up to the witness stand, and I'm dressed just like Dietrich in *Witness for the Prosecution*—stacked heels, black dress, simple pearls, black veil. The defense tries to trip me up, but they can't do it. I hold a handkerchief to my nose and I throw my veil over my forehead, and I point at Terry O'Sullivan, and I scream *She did it! She did it! I saw her do it!* And they have to *drag* me out of that courtroom. . . ."

During the Prince's speech, Clarisse had been thinking with furrowed brow. Then she said, "Terry O'Sullivan couldn't have killed Jeff King."

"Why not?" said the Prince. "When Terry O'Sullivan first came to the house he was wearing plaid pants. A man who wears plaid pants could be guilty of any crime."

"He couldn't have done it," said Clarisse. "Axel, where did you find Jeff King?"

Axel named a guesthouse to the east of the Throne and Scepter. "On the beach right behind it."

"And that's where Jeff and Terry had their fight?"

Axel nodded.

"Daniel, I thought Terry was with you the whole night," said Noah. "How did he get outside to *have* a fight with Jeff?"

"He went out with Ann and Margaret," said Valentine, turning to Margaret.

"Ann had had too much to drink. Terry said he'd help me get her home, but as soon as he had helped me out the door he split and ran down to the beach. Luckily I found a taxi. How long was Terry gone?" she asked Valentine.

"I was out with Clarisse on the deck then. So however long that was—twenty, thirty minutes."

Clarisse shook her head. "Which means that in twenty or thirty minutes, Terry would have had to have gone out, located Jeff, have the fight, make up, walk half a mile down the beach with him, kill him, cover him up with seaweed, and then get back to the party by the time Val and I had finished our cigarettes."

"But if he had hurried," said Noah. "Or if you are wrong about the time . . ."

"Another thing," said Clarisse. "When Terry and Val and I left the party, we walked directly to where the corpse was. Terry and Val left me right before I found it. If Terry had known that Jeff's body was there he would *never* have let us walk along the beach. I'm sure he didn't do it."

"You're right," said Scott, "I *know* he didn't do it."

All heads turned again.

"How do you know?" said Axel sharply.

Scott smiled. "He got his key back and then he went back toward the Crown."

"You were watching?" said Axel with some astonishment.

"If there's going to be a brawl," said the White Prince, "I'm leaving—unless somebody orders me another drink."

Valentine signaled for another round.

Scott continued, smiling unpleasantly at Axel as he spoke: "After our little run-in with Jeff King on the deck I went to the men's room, and when I came out again I saw you looking around for me. I saw you go outside, so I followed. If you had turned around you would have seen me. You went right down to the beach. And so did I. I was under the pier."

"God," said Axel with disgust. "If you're not being an exhibitionist, you're being a voyeur."

"No, just curious. I watched Terry O'Sullivan come down hard on you and Jeff, I watched you leave, and—" He paused for effect.

"And what!" demanded Clarisse.

"—and I watched Terry leave too, with the room key in his pocket."

"And Jeff King was left alone?" said Valentine.

"For about twenty seconds," said Scott. "Then somebody else came up."

"And who was that?" said Clarisse anxiously.

"I don't know his name," said Scott. "But it was the same man I saw snorting coke with Terry O'Sullivan at Back Street the night he died."

"You said before that you didn't see who Terry was with in the john," said Clarisse. "Come on, Scott, get your story straight!"

"I didn't see *all* of him," said Scott, "but I did see the top of his head."

"And from that you could identify him?" said Clarisse.

"Yes," said Scott.

The table was absolutely still for a moment. Then Valentine stood, looked hard at Scott, and then without warning snatched off Evita Piranha's wig. The Prince's white hair shone in the dim light of the bar. "Was it like this?" Valentine asked.

"Exactly like that," said Scott quietly.

The White Prince shrieked, swept his long arm across the table, and sent all the glasses, empty beer bottles, ashtrays, and candles crashing to the floor. He stood, and threw his hands over his face. He kicked over his chair. Angel ducked to one side and fell off her chair toward Noah, who leapt out of her way. The Prince turned to flee. Clarisse clutched at his skirt, but the Prince slapped her sharply across the face. She tumbled off her chair and tripped up Valentine, who had reached for the Prince's arm. The Prince smashed his fist so hard against Valentine's nose that blood spurted across the bodice of the Prince's dress and he howled in anger. Valentine fell on top of Clarisse. The Prince started to run, but Clarisse grabbed the heel of his shoe. He fell forward and crashed into the next table. The table collapsed and the Prince sprawled on the floor at the feet of several horrified onlookers. He plucked off his other shoe and flung it at Clarisse's head. It hit Valentine instead.

The crowd in the bar had seen only the confusion surrounding the White Prince. "Evita, Evita!" they cried.

The White Prince wrenched himself to his feet but stopped dead. The barrel of Matteo Montalvo's revolver was pointed directly at his chest.

Epilogue

Valentine and Clarisse sat at the kitchen table the next morning, silent, still, and nursing cups of steaming coffee. It was only half past eight, but already the day was hot and bright. Outside, in a brief bathing suit, Axel stood idly beside the pool, watching it fill with water. Noah was weeding the flower bed just outside their window.

Experimentally, Clarisse smiled. Then she grimaced, touching the large bruise on her right cheek.

"Don't complain," said Valentine. "I'm worse off than you are." He crossed his eyes to look down at the large strip of gauze and tape over his nose. Both his cheeks bore large circles of bruised flesh. "I'll bet the doctor set it wrong," he said. "When the tape comes off I'll have a cauliflower nose."

"Stop worrying. I've been offering up prayers to Debbie, goddess of nose jobs."

The screen door scraped open. Noah came in with dirty trowel in one hand and a bouquet of orange zinnias in the other. He bowed and handed the flowers to Clarisse. "The Prince may have been Queen of the Ball last night, but this morning you're certainly the Woman of the Hour."

"You're awfully chipper," said Valentine glumly to Noah. "Shouldn't you be in mourning or something?"

"The Prince gave me so much grief in the past five years that I don't have any sympathy left for him. I guess I'm just a hard woman."

"No, you're not," said Clarisse. "The Prince murdered Jeff King. Deliberately. He doesn't deserve any sympathy."

"We were at the doctor's all night," said Valentine. "You were at the police station. Did you find out *why* the Prince killed him?"

Noah replied calmly, "Victor loves coke. He was hooked on it—or he says he was hooked on it. Do you know how expensive that can be?"

Neither Valentine nor Clarisse bothered to answer so obvious a question.

Noah crossed his arms. "If I showed you the books at the restaurant and lent you an adding machine, you could find out exactly how much it costs."

"The Prince was playing the piano?" said Valentine surprised. "He was robbing the till?"

"For the past year and a half," replied Noah. "Ever since he began working there, in fact. Siphoning off a little every time he bought a couple of grams. It's one of the reasons I wanted him to leave this season."

"Why didn't you just tell him to stop?" asked Clarisse.

Noah shrugged. "Why didn't I kick Jeff King out before he robbed me blind? I guess I'm no good at endings. I'm not very good at beginnings either. I'm great at middles, though. Anyway, Angel figured out what was going on, and she came to me and I said, 'Don't do anything, I'll replace the money,' and that's what I did."

"But how did Jeff King figure in all this?"

"Jeff figured out what was happening, I don't know how. Maybe he just put two and two together. The Prince was buying all this coke, and his job didn't pay *that* much. But what I think happened is that he found out about it from one of the waiters at the Swiss Miss—Jeff supplied a lot of people in town. Anyway, he doubled his price on the coke he sold the Prince and told him if he didn't pay he'd come straight to me."

"I think of that sweet young man on the ferry," mused Clarisse. "Drug pusher, blackmailer, petty thief, co-respondent in homosexual divorce cases, murder victim . . ."

"That Saturday afternoon when Jeff came by here, the Prince saw him," said Noah.

"The Prince wasn't here then," said Clarisse.

"No, but he did see Jeff leave, and he figured that Jeff had come to tell me about the cooked books. But when I didn't say anything, he realized that he had a grace period."

"The Prince didn't know you knew about the books?" said Valentine.

"He had no idea," sighed Noah. "What a dummy. In a restaurant everybody knows everything."

"So he planned to kill Jeff that night?"

"I don't know," said Noah. "The Prince spent last night gulping out a confession between sobs, but he didn't tell everything."

"How'd you get to hear all this?" asked Valentine.

"I sat outside the room where they questioned him—I more or less had my ear against the door. The Prince wouldn't shut up. Anyway, after we came back here from the party, he got out of his Salome drag and went for a walk. I don't think he was looking for Jeff, but he sure enough found him. So they did some coke together and walked along the beach, and when they got to a place where the Prince couldn't see anybody, the Prince took off one of his sandals—you know, the wooden ones—and hit Jeff over the head with it. That stunned him and then the Prince strangled him."

"Then covered him up with seaweed," Clarisse put in. "But did the Prince also kill Terry O'Sullivan?"

"Well, he gave him the coke in the ladies' room at Back Street," said Noah, "but he didn't know Terry had a heart condition. The Prince forgot to latch the stall. Terry O'Sullivan barged in, looped to the gills, and saw the Prince doing coke and started going on and on about how gay people shouldn't do hard drugs. The Prince told him to mind his own business, and then Terry said he was going to turn him in to the cop on the door. So the Prince said, 'Hey! Coke's not really hard drugs, try some.' And he gave Terry four good lines and Terry went out and dropped dead at your feet. It was an accident."

"Two down," said Valentine. "Now the question is, did Terry O'Sullivan kill Ann Richardson?"

"That I don't know," said Noah. "Victor confessed to everything but that and the kidnapping of the Lindbergh baby."

"I still think Terry O'Sullivan killed Ann," said Clarisse.

"I do too," agreed Valentine, "but did he do it on purpose?"

"When I went through the autopsy reports day before yesterday, I realized that Ann didn't have any bruises on her," said Clarisse. "It's pretty hard to drown somebody, especially somebody who could swim as well as Ann could. There would have been a struggle."

"But he might have mixed the drugs on purpose. He probably thought she would say something to someone that would implicate him in the murder of Jeff King."

"Pretty bad reason to murder someone, particularly since Terry *didn't* kill Jeff. I vote that it happened by accident, just the way he told you it did. But afterward he was afraid to go to the police and tell them he was the one who provided the drugs."

"I've got a question," said Axel, who had been standing outside the door listening.

"What?" said Valentine.

"How'd you know to bring everybody together last night? If all those people hadn't been there, you wouldn't have found out anything at all."

Clarisse preened, and smiled despite the pain it caused her. "That was my doing. I knew all the victims, so I figured I knew the murderer too. So I just sent out invitations."

"It wouldn't have worked without Margaret being there," Noah pointed out. "So there was at least a bit of luck involved."

"Did you suspect the Prince?" asked Axel.

"We weren't sure who did it," Valentine said.

Clarisse nodded agreement.

"But who *did* you think did it?" said Axel.

Now Valentine looked embarrassed.

Clarisse spoke up. "We were hoping it was Scott."

Axel laughed. "And if it had been, that would have broken us up, right?"

"Right," said Clarisse. "It was none of our business, though."

"Well," said Axel with a smile, "you will be happy to hear that I have cut the cord. Last night while all of you were at the doctor's and at the police station, Scott and I came back here.

I told him I didn't like being spied on and that I didn't like to be made to feel guilty for playing around and that I was tired of supporting him and that I wanted him out of my life."

"Well! How'd he take it?" asked Valentine.

"Pretty badly. I had always been the one who transgressed, and he was always the one who forgave. Now he wanted *me* to forgive, but I wouldn't do it. He's back in Plymouth now, packing up. And today I start looking for my future ex-husband."

Valentine grinned, stretching the adhesive on his nose. The telephone rang, and he went into the next room to answer it.

"Do you think I'll get in the paper?" Clarisse asked her uncle.

"I wouldn't count on it," Noah said doubtfully. "The photographers were there last night, and there's probably going to be a photograph of the White Prince—as Eva Perón—clutching the bars and glaring. But the reporters didn't ask many questions. I guess they'll just take away copies of the confession or something."

"Am I mentioned in the confession?" asked Clarisse.

"No," said Noah, gently. "I don't remember his mentioning you."

"Oh, God," sighed Clarisse. "At least I'll get to testify at the trial."

"He confessed," said Axel. "There won't be any trial. Just a sentencing."

Clarisse beat her fists on the table. "God! I have lived this entire month in complete and total obscurity. I tell you, from here on out this summer had better get itself into shape! I want perfection! I want love and a tan! I want to receive a vast amount of money for doing nothing at all!"

Valentine came back in.

"Was that a reporter asking to set up an interview?" asked Clarisse hopefully.

"No," said Valentine. "It was Angel. She was angry."

"What about?" asked Noah surprised.

"About the damage the Prince did last night. The repairs have to come out of the door receipts. So I said we'd make it up to her . . . somehow."

Clarisse detected the evasion in his voice. *"How?"* she asked darkly. "How are we going to make it up to her?"

200

"Well," said Valentine. "You and I have just been entered in the July Fourth Marathon Kickline to End World Hunger."

"I should have guessed," breathed Clarisse. "Oh, God," she sighed, "as I remarked to Jeff King on the day he died, I *hate* Provincetown."

The long-awaited new novel by the author
of GAYWYCK. A sensitive, emotional novel
of the fragile relationship between a
successful novelist and a handsome poet,
and the destructive effects of alcoholism
on the two men.

A COMFORTABLE CORNER
Vincent Virga

Terence, a well-known novelist, has just
separated from his lover, Christopher, a popular poet whose losing battle with liquor is wors-
ened by his own father's alcoholism. Though
Terence's love for Christopher is deep, he
knows that only Christopher can save himself.
As they drift into new affairs, each reflects
upon their past love together as the only
meaningful one in their lives. But only when
Christopher's father meets an untimely al-
cohol-related death does he realize the
depths to which he has fallen...enabling him
to seek professional help and return to Terence for another try at love.